Engineering Applications
of Pneumatics and Hydraulics

Engineering Applications of Pneumatics and Hydraulics

by EurIng Ian C. Turner

BSc CEng FIPlantE MIEE FInstMC MCIBSE
Recipient of the '1997 PLANT ENGINEER OF THE YEAR' Award

In collaboration with

The Institution of Plant Engineers

BUTTERWORTH
HEINEMANN

OXFORD AMSTERDAM BOSTON LONDON NEW YORK PARIS
SAN DIEGO SAN FRANCISCO SINGAPORE SYDNEY TOKYO

Butterworth-Heinemann
An imprint of Elsevier Science
Linacre House, Jordan Hill, Oxford OX2 8DP
225 Wildwood Avenue, Woburn MA 01801-2041

First published by Arnold 1996
Reprinted by Butterworth-Heinemann 2001, 2002

British Library Cataloguing in Publication Data
A catalogue record for this book is available from the British Library

Library of Congress Cataloging-in-Publication Data
A catalogue record for this book is available from the Library of Congress

ISBN 0 340 62526 0

For information on all Butterworth-Heinemann publications
visit our website at www.bh.com

Typeset by GreenGate Publishing Services, Tonbridge, Kent
Printed and bound in Great Britain by J W Arrowsmith Ltd, Bristol

Contents

Preface

Fluid power systems technology has been with us for many years and it is a form of technology which is likely to remain with us for the foreseeable future. However, like all technology there have been dramatic changes over the years, most notably in the materials used in the construction of pneumatic and hydraulic components, the arrangement of circuits and the speed and reliability of operation of pneumatic and hydraulic systems.

This text is based upon a series of training modules developed for short courses in Basic Pneumatics and Air Logic which have been delivered over recent years as in-company training material to a number of end user organisations over a diverse range of industries, from petro-chemicals to automobile manufacture to food processing.

Whilst initially the training material was developed for short course delivery it has now been expanded to cater for formal qualification programmes and modules in Engineering as offered by a number of training providers. In this respect the text is now suitable for relevant City & Guilds and BTEC modules/programmes of study in colleges of FE and HE whilst at the same time catering for the updating needs of practising engineers. The book is also designed to cater for students engaged upon N/SVQ and GN/SVQ programmes of study in Engineering by adequately covering the unit range for fluid power in both mandatory and optional units.

The material has been presented in a practical, readable format and is well illustrated with diagrams of components and circuits as used in everyday pneumatic and hydraulic applications. It is the ideal text to supplement both work and college based study of this important area of technology.

Ian C. Turner

Acknowledgements

The author expresses grateful thanks to the following organisations for their support and kind permission to reproduce extracts from their training material and other publications:

- IMI Norgren
- Festo Didactic KG
- Brisco Engineering
- Spirax Sarco
- Webtec Hydraulics
- City & Guilds of London Institute
- Institution of Plant Engineers.

The Institution of Plant Engineers

Plant engineers are the people responsible for the efficient operation and maintenance of fixed and mobile plant, and allied works services. The Institution is their natural home, providing them with a range of services and publications, branch activities, career guidance and, where appropriate, Engineering Council registration. They can be contacted at 77 Great Peter Street, Westminster, London SW1P 2EZ. Tel: 0171 233 2855 Fax: 0171 233 2604.

1 Applications of Pneumatics and Hydraulics in Industry

Aims

At the end of this chapter you should be able to:

1 Appreciate a range of industrial applications for pneumatics and hydraulics.
2 Appreciate that pneumatics and hydraulics may be used in combination with other technologies in a given system.
3 Recognise that fluid power systems may be used for operating, controlling and/or taking measurements of equipment, machinery and plant.
4 Have an awareness that fluid power systems can be used in industrial processes requiring emergency and safety shut-down arrangements.

1.1 Industrial applications

Pneumatic and hydraulic systems have been used for many years within industrial processes and as such have acquired an established place in modern industry. Continuous development of fluid power technology over the years has significantly expanded and increased the applications to many areas hitherto not known for adopting pneumatic and hydraulic technology.

Some of the principal users of fluid power technology are:

- **manufacturing industries**, notably the automotive industry, machine tool manufacturers and domestic and commercial appliance manufacturers
- **processing industries**, such as chemical, petro-chemical, food processing, textiles, paper, etc.
- **transportation systems**, including marine and mobile construction plant
- **utilities**, particularly in the gas industry
- **defence systems**.

More recent users have been in the fields of offshore oil and gas development, space and aeronautical systems and nuclear applications.

1.2 Combined technologies

Often, pneumatics and hydraulics are combined with other technologies such as mechanical, electrical and electronic systems to form an overall system. An example of this can be found in **robotics**.

In addition, safety conscious industries will sometimes adopt a number of technologies operating on different physical principles as a means of achieving diversity of operation, control and measurement on a given process. This is particularly significant as protection against common mode failure whereby if one system fails, the others remain active.

1.3 Uses of fluid power systems

Fluid power systems may be used for:

1 **Carrying out work by operating plant and machinery using linear, swivel and rotary motion**. Some general methods of material handling used in industry, for example, may be:

- clamping
- shifting
- positioning
- orientating.

a) General applications may be:

- packaging
- feeding
- door or chute control
- material transfer
- turning and inverting of parts
- sorting
- stacking
- stamping and embossing.

Figure 1.1 A combined wellhead control panel and hydraulic power unit for an offshore oil platform

b) Some general machining and work operations may be:

- drilling
- turning
- milling
- sawing
- finishing and buffing
- forming.

2 **Controlling processes and plant**. Pneumatic and hydraulic systems may be used to sense the operational status of a process, feed this information back to a controller which will take a necessary control action, for example a limit switch may sense that an actuator needs to be operated.

3 **Measurements of process and/or machine parameters**. Pneumatics and hydraulics can be used to provide measurements of process or machine parameters, act on this information and subsequently display it to an operator.

The processes outlined in 1, 2 and 3 above may be used individually or in combination.

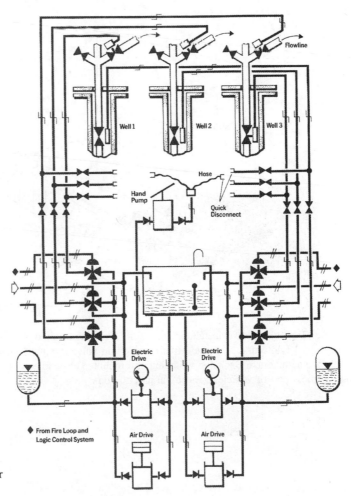

Figure 1.2 Practical hydraulic emergency shut-down system for three sub-sea oil wells

1.4 Hydraulic and pneumatic safety systems

In addition to operating, controlling and measuring parameters of process plant and machinery, hydraulics and pneumatics may be used in high integrity safety systems. This is expanded further in Chapter 14.

The high speed and reliability of operation embodied in good modern pneumatic and hydraulic system design coupled with inherent explosion-proof and overload-safe operation makes the choice of this technology ideal for applications in the marine, offshore and petro-chemical industries.

Figure 1.2 shows a practical hydraulic emergency shut-down system for three sub-sea oil wells on an offshore petro-chemical application.

A further example of a safety system is that used in automatic reverse braking systems which can be fitted to any vehicle using air or air/hydraulic brakes. Such a system can bring a vehicle to a halt literally within centimetres once an obstruction is encountered. Figure 1.3 shows a typical arrangement of such a backstop system.

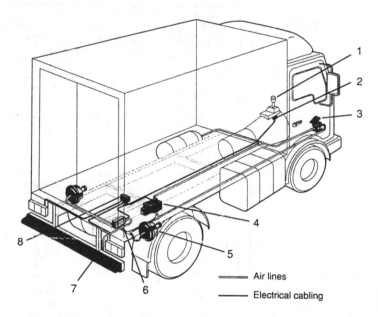

====== Air lines
———— Electrical cabling

Figure 1.3 Typical backstop installation on a fully air braked rigid truck. 1 Gear change selector, 2 Reversing light switch, 3 Footbrake, 4 Brake actuating solenoid valve mounted adjacent to rear brake supply, 5 Rear spring brake actuators, 6 Backstop switching unit mounted close to rear of vehicle, 7 Backstop sensor mounted on rearmost part of vehicle, 8 Electrical junction box

2 Basic Principles of Fluid Power Systems

Aims

At the end of this chapter you should be able to:

1 Appreciate the nature and physical properties of air.
2 Understand that a fluid power system is a compressed gas or incompressible liquid operating under enclosed conditions in order to produce work.
3 Recognise the SI system of units of measurement for use in pneumatics and hydraulics.
4 Be aware of the physical laws governing gases, namely Newton's and Boyle's laws.

2.1 Physical properties of air

The surface of the Earth is covered entirely by layers of air. It is an abundant gas mixture comprising:

- nitrogen (approximately 78% volume)
- oxygen (approximately 21% volume).

There are also small amounts of water vapour and carbon dioxide together with traces of the 'rare gases': argon, krypton, xenon, neon and helium.

When air is compressed and stored it can be used as a medium for making measurements and for controlling and operating equipment and plant. This is known as **pneumatics**.

2.2 The principle of hydraulic systems

In hydraulic systems, the compressed air is replaced with a liquid-based solution, typically oil, which is used under pressure to perform measurements and to control and operate plant and machinery. Hydraulic systems tend to operate at much higher pressures than pneumatic systems and consequently can produce larger forces.

Over the years the performance standards of hydraulic equipment have risen. Whereas a pressure of about 70 bar used to be adequate for industrial hydraulic systems, nowadays systems operating with pressures of 150 to 250 bar are common. In certain applications pressures in excess of 350 bar are used, for example large industrial presses, offshore petro-chemical installations, etc.

2.3 The fluid power system

The behaviour of both compressed gases and incompressible liquids are similar when operating under enclosed conditions. Gases and liquids are considered to be fluids and can be used to transmit energy over long distances. Such systems are commonly known as **fluid power systems**.

2.4 SI system of units

To assist in the understanding of the natural laws of fluid power as well as its behaviour, physical dimensions are employed and systems of units of measurement for use in pneumatics and hydraulics have been developed. The system that is commonly used nowadays is called the Systéme International d'Unités or SI system. These units are based on metric units and the following terms and units are generally used:

Base quantities

Unit	Symbol	SI System
Length	L	metre (m)
Mass	m	kilogram (kg)
Time	t	second (s)
Temperature	T	kelvin (K)

Derived quantities

Unit	Symbol	SI System
Force	F	newton (N)
Area	A	square metre (m^2)
Volume	V	cubic metre (m^3) $1\ m^3 = 1000$ litres
Flow rate	Q	cubic metre per second (m^3/s) 1 litre/s $= 10^{-3}\ m^3/s$
Pressure	P	pascal (Pa) $1\ Pa = 1\ N/m^2$ $1\ bar = 10^5\ Pa = 100\ kPa$

Pressure

Pressure is a force acting at right-angles onto a surface area. One (1) pascal represents a constant pressure on a surface area of 1 m² with a force of 1 N (newton) acting at right-angles to that surface area (Figure 2.1). 10^5 Pa or 100 kPa is equal to 1 bar and this represents atmospheric air pressure (which is 14.5 psi in the old imperial system of units). As can be seen, the pascal is a very small unit of pressure and hence we tend to use the **bar**. **One (1) bar represents atmospheric pressure**.

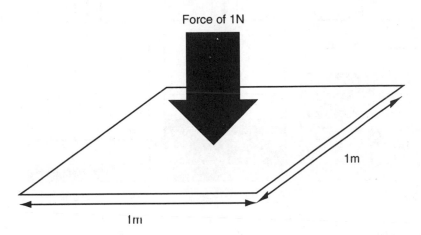

Force of 1N

1m

1m

Figure 2.1 Pressure of one pascal

2.5 Newton's law

Air is a mass of gases. Equally, in the case of hydraulics, oil is a mass of liquid. Both can be related to force from Newton's law:

Force = mass × acceleration

$F = ma$

where

a is replaced by the acceleration due to gravity, g:

$g = 9.81$ m/s².

2.6 Boyle's law

In common with all gases, air has no particular shape. Its shape changes with the slightest resistance, that is it assumes the shape of its surroundings or container. As we know, air can be compressed, in which case it endeavours to expand. The applicable relationship between pressure and volume of containment of a gas is given by **Boyle's law**.

Boyle's law states:

At constant temperature, the **volume** of a given **mass** of gas is **inversely proportional** to the **absolute pressure**. In other words the product of **absolute pressure** and **volume** is **constant** for a given **mass** of gas (Figure 2.2).

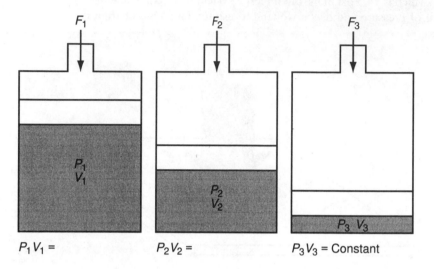

$$P_1 V_1 = \qquad P_2 V_2 = \qquad P_3 V_3 = \text{Constant}$$

Figure 2.2 Boyle's law

The liquids used in hydraulic systems are generally considered to be incompressible and insensitive to changes in temperature, unlike the case of pneumatics where air is compressed and is very sensitive to changes in temperature. In this respect, additional plant and equipment are installed on compressed-air installations in order to control temperature changes resulting from compressibility, for example air coolers and heaters. This is discussed further in Chapter 5.

Questions

1 What are the two gases that mainly make up air? State their approximate percentage volume.
2 State the SI units that are used for measurements of:
   (a) force
   (b) area
   (c) volume
   (d) flow rate
   (e) pressure.
3 What is the value of pressure (in bar) resulting when a force of 100 000 newtons acts upon an area of 1 square metre?
4 If the absolute pressure of 2 m³ of air is increased from 2.4 bar to 2.75 bar, find the new volume, given that the temperature remains constant.

3 Features and Characteristics of Pneumatic and Hydraulic Systems

Aims

At the end of this chapter you should be able to:

1 *Distinguish between atmospheric and gauge pressure.*
2 *Appreciate some of the advantages and distinguishing characteristics of compressed air and hydraulic systems.*
3 *Appreciate some of the disadvantages and limitations of pneumatics and hydraulics.*
4 *Recognise the basic requirements for mains air supplies.*

3.1 Air pressure relationship – atmospheric and gauge pressure

Since everything on Earth is subjected to the absolute atmospheric pressure (P_{at}), this pressure cannot be felt. The prevailing atmospheric pressure is therefore regarded as the base and any deviation is termed:

- gauge pressure = P_g, or
- vacuum = P_v.

The air pressure relationship is illustrated in Figure 3.1. The atmospheric pressure does not have a constant value, it varies with geographical location and weather. The range from the absolute zero line to the variable atmospheric pressure line is called the vacuum range and above this, the pressure range. The absolute pressure P_{ab} is composed of atmospheric pressure P_{at} and gauge pressure P_g. In practice, gauges are used which show only the excess pressure P_g.
P_{ab} is approximately one bar (100 kPa) greater than the P_g value.

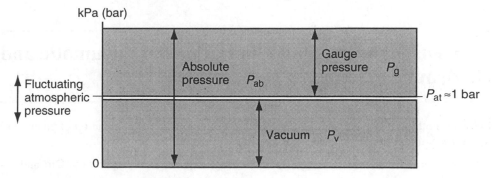

Figure 3.1 Measuring pressure and vacuum

3.2 Advantages and distinguishing characteristics of compressed air and hydraulic systems

As we have seen in Chapter 1, pneumatics and hydraulics have for some time been used in industry for measurement, control and operational applications, for example:

- employing the use of sensors to determine the operational status of processes
- information and data processing
- the physical operation of actuators
- carrying out work.

The principal advantages and distinguishing characteristics of pneumatics and hydraulics are listed below.

- **Availability.** Air is available practically everywhere in abundant quantities.
- **Transport.** Air and hydraulic fluid can be transported easily in pipelines, even over large distances.
- **Storage.** An air compressor or hydraulic pump need not be in continuous operation. Compressed air can be stored in a reservoir and removed as required. Similarly, hydraulic fluid can be stored in accumulators.
- **Temperature.** Hydraulic fluids are relatively insensitive to temperature fluctuations. This ensures reliable operation, even under extreme conditions.
- **Explosion proof.** Compressed air and hydraulic fluid offer minimal risk of explosion or fire, hence no expensive protection against explosion is required.
- **Cleanliness.** Unlubricated exhaust air is clean. Any unlubricated air which escapes through leaking pipes or components does not cause contamination. This is an important point when considering the food, timber/woodwork and textile industries.
- **Components.** The operating components for pneumatics and hydraulics are of simple construction and are therefore relatively inexpensive.
- **Speed.** Compressed air and hydraulic fluid are very fast-working media. This enables high working speeds to be attained.
- **Adjustable.** With compressed air and hydraulic components, speeds and forces are variable over wide ranges.
- **Overload safe.** Pneumatic and hydraulic tools, machinery and operating components can be loaded to the point of stopping and are therefore overload safe.

3.3 Disadvantages and limitations of pneumatic and hydraulic systems

Obviously nothing is perfect and there are certain limitations in the application of pneumatics and hydraulics. Some of these negative characteristics are listed below.

- **Preparation.** Compressed air and hydraulic fluid require good preparation. Dirt and condensate should not be present.
- **Compressible.** Compressed air does not always achieve uniform and constant piston speeds.

- **Force requirement.** Compressed air is economical only up to a certain force requirement. Under the normal working pressure of 6–7 bar (600 to 700 kPa) and dependent on the travel and speed, the output limit is between 20 000 and 30 000 Newtons.
- **Noise level.** In pneumatic systems exhaust air is loud. This problem has now, however, been largely solved due to the development of sound absorption material and silencers.
- **Costs.** Compressed air and hydraulic media are relatively expensive means of conveying power. The high energy costs are partially offset by inexpensive components and higher performance.

3.4 Basic requirements for mains air supplies

It is essential for proper and effective operation in all pneumatic systems that mains air supplies be:

- clean
- dry
- oil free.

If air becomes contaminated with dirt, high concentrates of moisture and oil etc. then this will have an adverse effect upon the operation of pneumatic components and equipment, for example valve/actuator seizure and sticking.

In practice, equipment and plant are provided to generate, distribute and supply clean, dry and oil-free air. This will be looked at in more detail in Chapter 5.

Questions

1 Complete the following table:

Absolute pressure (P_{ab})(bar)	Gauge pressure (P_g)(bar)
6	–
–	2.5
17	–
–	8.9

2 A compressed air storage system comprises six 0.25 m³ air bottles giving a total volume of 1.5 m³. During operation the system undergoes a change in gauge pressure from 9.2 bar to 5.8 bar. With the temperature remaining constant and assuming each air bottle is exhausted in turn, what is the remaining total volume and number of bottles under pressure?

3 Compare and contrast advantages and disadvantages of compressed air and hydraulics for an application requiring operation in a Zone 1 hazardous area, requiring fast working speeds of operation and low ambient noise levels.

4 Component, Equipment and Plant Symbols

Aims

At the end of this chapter you should be able to:

1 *Have an awareness of the relevant CETOP, International and British Standards relating to symbols used for pneumatic and hydraulic components, equipment and plant.*
2 *Recognise certain common symbols used in pneumatic and hydraulic circuit diagrams.*
3 *Understand component lettering and numbering identification systems.*
4 *Relate the use of lettering and number designations to component selection.*

4.1 Identification of graphical symbols used in pneumatics and hydraulics

During 1964 the Comité Européen des Transmissions Oleohydrauliques et Pneumatiques (CETOP) published a proposed range of symbols for hydraulic and pneumatic equipment. These symbols were later adopted by the International Standards Organisation in its document ISO 1219 and this in turn subsequently formed the basis of the British Standard **BS 2917 (1977) – Graphical symbols used on diagrams for fluid power systems and components**.

Appendix 4 lists some international fluid power standards together with relevant standardisation organisations.

There are some differences between CETOP, ISO and BS symbols, but all are intended to be self explanatory. Also, it should be noted that sometimes manufacturers produce a piece of pneumatic or hydraulic equipment which cannot be exactly represented by existing symbols, in which case the manufacturer develops its own symbol. Thus companies such as Festo, IMI Norgren, Bosch, etc. may produce their own symbols for particular equipment and components.

The following shows the more commonly applied BS 2917 symbols. A more detailed description of the symbols relating to fluid power generation, distribution, supply equipment, valves and actuation will be given in later chapters.

4.2 Energy conversion symbols

Note: For hydraulic components the part symbol ∇ will be filled in thus ▼ e.g. A hydraulic motor would be shown as

Compressor

Pneumatic motor, constant, with one direction flow

Pneumatic motor, adjustable, with one direction flow

Pneumatic semi-rotary actuator, with limited angle of rotation (rotary cylinder), with two directions of rotation

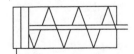

Single-acting cylinder, return stroke by spring

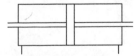

Double-acting cylinder with double-ended piston rod

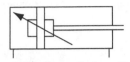

Double-acting cylinder with cushioning adjustable at both ends

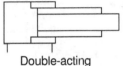

Double-acting telescopic cylinder

Pneumatic-hydraulic rodless actuator

Vacuum pump

Pneumatic motor, constant, with two direction flow

Pneumatic motor, adjustable, with two direction flow

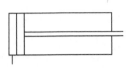

Single-acting cylinder, return stroke by external force

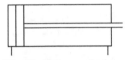

Double-acting cylinder

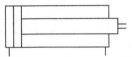

Differential cylinder

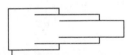

Single-acting telescopic cylinder

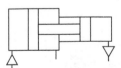

Pressure intensifier

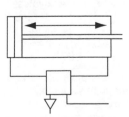

Pneumatic linear unit

4.3 Valve symbols

Directional control valves

A (2)

(1)

2/2-way valve, closed normal position

(2) A

(1) P

2/2-way valve, open normal position

(2) A

(1) P R (3)

3/2-way valve, closed normal position

(2) A

(1) P R (3)

3/2-way valve, open normal position

(2)A

(1) P R (3)

3/3-way valve, closed neutral position

(4)A B(2)

(1)P R (3)

4/2-way valve

(4)A B(2)

(1) P R (3)

4/3-way valve, closed neutral position

(4)A B(2)

(1)P R (3)

4/3-way valve, neutral position, working lines vented

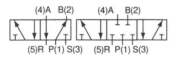
(4)A B(2)   (4)A   B(2)

(5)R P(1) S(3)   (5)R P(1) S(3)

5/2-way valve   5/3-way valve, closed neutral position

Direction control valve with intermediate switching positions and two final positions

Flow control valves

Throttle valve with constant restriction

Diaphragm valve with constant restriction

Throttle valve, adjustable, any type of operation

Throttle valve, adjustable, mechanical operation against return spring

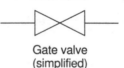
Throttle valve, adjustable, manual operation

Gate valve

Gate valve (simplified)

Special symbols*

P
X
Reflex sensor

X
P
Back pressure nozzle

P
Nozzle, general, emitter nozzle for air gate

P
X
Collector nozzle, with air supply for air gate

Pressure control valves

P (1)

R (3)

Pressure relief valve, adjustable

P (1)

A (2)

Sequence valve, adjustable, with pressure relief

P (1)

A (2)

Pressure regulator, adjustable

(1) P R (3)

A (2)

Relieving pressure regulator, adjustable

Non-return valves

Check valve without spring

Check valve with spring

Pilot-controlled check valve

A
X   Y
Shuttle valve

(2) A
(1) P   R
Quick exhaust valve

One-way flow control valve, adjustable

Two-pressure valve

4.4 Energy transmission symbols

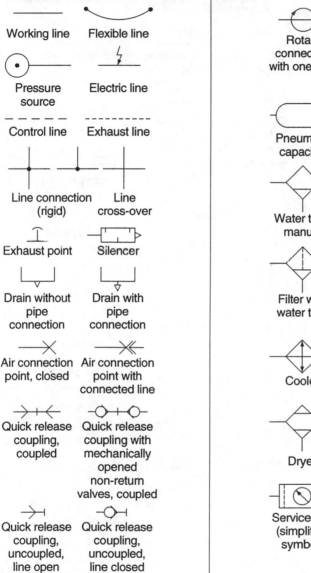

Working line   Flexible line

Pressure source   Electric line

Control line   Exhaust line

Line connection (rigid)   Line cross-over

Exhaust point   Silencer

Drain without pipe connection   Drain with pipe connection

Air connection point, closed   Air connection point with connected line

Quick release coupling, coupled   Quick release coupling with mechanically opened non-return valves, coupled

Quick release coupling, uncoupled, line open   Quick release coupling, uncoupled, line closed

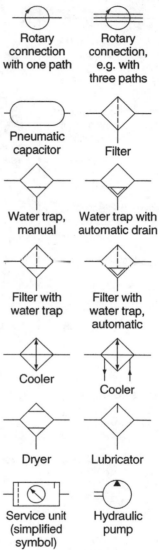

Rotary connection with one path   Rotary connection, e.g. with three paths

Pneumatic capacitor   Filter

Water trap, manual   Water trap with automatic drain

Filter with water trap   Filter with water trap, automatic

Cooler   Cooler

Dryer   Lubricator

Service unit (simplified symbol)   Hydraulic pump

4.5 Control symbols

Mechanical components

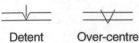

Detent   Over-centre
              device

Shaft rotational Shaft rotational
movement in   movement in
one direction   two directions

Locking device   Hinge joint
(*control   with extended
method for   lever
releasing the
locking device)

Hinge joint   Hinge with
simple   fixed pivot

Pressure controls

Direct by   Pressure
application of   centred
pressure

Differential   Spring
pressure   centred
actuation

Indirect by application
of pressure

Combined controls

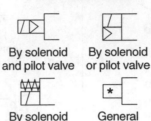

By solenoid   By solenoid
and pilot valve or pilot valve

By solenoid   General
or manual   *explanatory
operation with   symbol
return spring

Manual controls

General   Button

Lever   Pedal

Mechanical controls

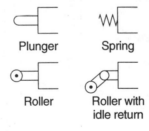

Plunger   Spring

Roller   Roller with
              idle return

Electrical controls

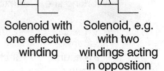

Solenoid with   Solenoid, e.g.
one effective   with two
winding   windings acting
                in opposition

Electric motor   Electric
with continuous stepping
rotary motion   motor

Special controls

By application   By application
of pressure   of pressure,
through   type of control
pressurised   produces
amplifier*   alternating
                  behaviour*

* = not standardised, proposal

Mechanical   Actuated in
return   starting position

4.6 Other symbols

Other devices

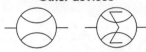

Flow
measuring
instrument
(flow)

Flow
measuring
instrument
(volume)

Temperature
sensor

Temperature
gauge

Pressure
gauge

Pressure
switch

ISO STANDARD 5599/II
Designation of Connections

A, B, C (2, 4, 6) working lines

P(1) compressed air connection

R, S, T (3, 5, 7) drain, exhaust points

L(9) leakage line

Z,Y,X(12,14,16) control lines

Logic Symbols

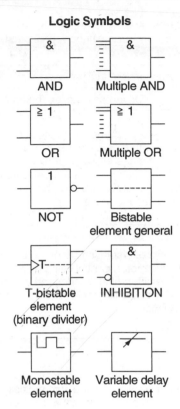

AND Multiple AND

OR Multiple OR

NOT Bistable element general

T-bistable element (binary divider) INHIBITION

Monostable element Variable delay element

4.7 Examples of assemblies of equipment

In circuit diagrams, symbols normally represent equipment in the unoperated condition. However, any other condition can be represented, if clearly stated.

Description and interpretation of the examples	Symbol
Driven assemblies (pumps)	
A two-stage pump driven by an electric motor with a pressure relief valve in the second stage and a proportioning relief valve which maintains the pressure of the first stage at, for example, half the pressure of the second stage.	
A variable displacement pump driven by an electric motor, control being by a servo-motor with a differential cylinder and a tracer valve, with two throttling orifices and mechanical feedback.	
A single stage air compressor driven by an electric motor which is automatically switched on and off as the receiver pressure falls and rises.	
A two stage air compressing assembly driven by an internal combustion motor which idles or takes up the load with the switching over of a 3/2 directional control valve, depending on the receiver pressure.	
Driving assemblies (motors)	
A motor driven in either direction of rotation, with pressure relief valves and flushing valve.	

Examples of assemblies of equipment (continued)

Control and regulating assemblies	
A control unit by which the piston of a cylinder is automatically moved back and forth.	
A group of two 6/3 directional control valves which are connected to separate non-return valves and to a common pressure relief valve. When both directional control valves are in the neutral position, the flow is returned to the reservoir.	

4.8 Examples of complete installations

In circuit diagrams, symbols normally represent equipment in the unoperated condition. However, any other condition can be represented, if clearly stated.

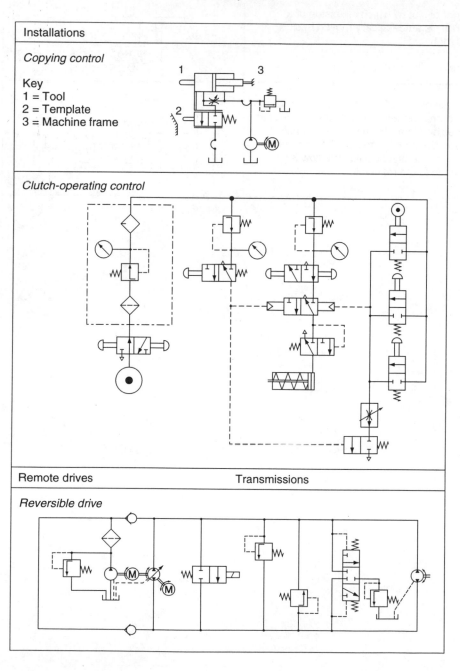

Installations

Copying control

Key
1 = Tool
2 = Template
3 = Machine frame

Clutch-operating control

Remote drives   Transmissions

Reversible drive

4.9 Component identification

Valve symbol description

In general the symbols are similar for pneumatic and hydraulics but each control medium has specific characteristics that are unique.

Valve switching positions are represented as squares		Shut off positions are identified in the boxes by lines drawn at right-angles	
The number of squares shows how many switching positions the valve has		The connections (inlet and outlet ports) are shown by lines on the outside of the box and are drawn in the initial position	
Lines indicate flow paths, arrows show the direction of flow			

Figure 4.1 Directional control valves – symbol development

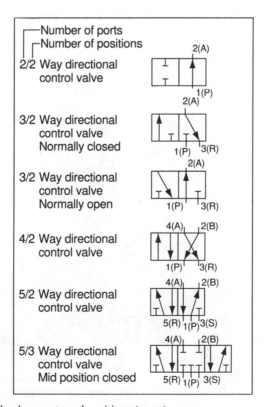

Figure 4.2 Directional control valves ports and positions (ways)

The directional control valve is represented by the number of controlled connections and the number of positions (Figure 4.1). Each position is shown as a separate square. The designation of the ports is important when interpreting the circuit symbols and the valve as fitted to the physical system. To ensure that the correct lines, connections and valves are physically in place, there must be a relationship between the circuit and the components used.

Therefore all symbols on the circuit must be designated and the components used should be labelled with the correct symbol and designations (Figure 4.2).

A numbering system is used to designate directional control valves and is in accordance with ISO 5599. Prior to this a lettering system was used. Both systems of designation are shown as follows:

Port or connection	ISO 5599 Numbering system	Lettering system
Pressure port	1	P
Exhaust port	3	R (3/2-way valve)
Exhaust ports	5, 3	R, S (5/2-way valve)
Signal outputs	2, 4	B, A
Pilot line opens flow 1 to 2	12	Z (single pilot 3/2-way valve)
Pilot line opens flow 1 to 2	12	Y (5/2-way valve)
Pilot line opens flow 1 to 4	14	Z (5/2-way valve)
Pilot line flow closed	10	Z, Y
Auxiliary pilot air	81, 91	Pz

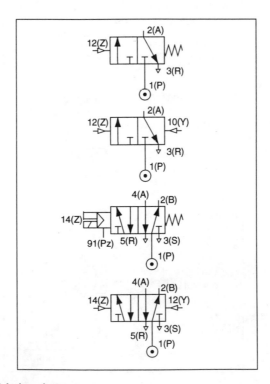

Figure 4.3 Examples of designations

Questions

1 Complete the grid below, either naming the symbol or drawing in the named symbol.

Lubricator		Silencer	
		Four-port valve two positions	
Solenoid operation			
		Two-port valve	
Double-acting non-cushioned cylinder			
Spring operation			

2 Sketch BS 2917 symbols for each of the following:
   (a) a 3/2-way directional control valve normally open with direct pneumatic actuation and spring return
   (b) a 5/2-way directional control valve with push-button actuation and spring return.

3 For each of the symbols shown below:
   (a) State the component giving a description of the device.
   (b) Ports, positions, method of actuation, etc. should be identified where appropriate.

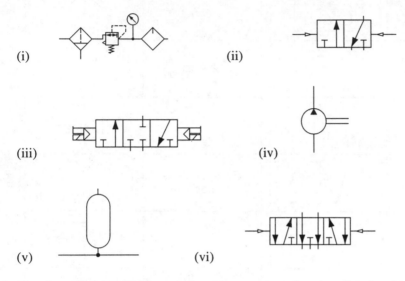

(i)   (ii)

(iii)   (iv)

(v)   (vi)

4 Sketch, using a BS 2917 symbol, a 2/2-way normally open directional control valve with foot pedal actuation and spring return.

5 Fluid Power Generation, Supply and Distribution

Aims

At the end of this chapter you should be able to:

1 Appreciate the arrangement and type of plant and equipment required for fluid power generation and supply.

2 Recognise appropriate generation and supply symbols on pneumatic and hydraulic circuit diagrams.

5.1 Air generation and distribution

We noted in Chapter 3 that it is essential for mains air supplies to be:

- clean
- dry
- oil free.

The preparation of the air starts from the point of generation. Contamination of the air can occur at many potential points in the generation and distribution system right up to the point of use. There is little point in preparing good quality air and then allowing incorrect component and equipment selection to reduce the quality. The equipment to be considered in the generation, preparation and distribution of air includes:

- compressor
- receiver
- filter
- dryer
- lubricator
- pressure regulator
- drainage points
- oil separators.

5.2 Compressor types

Positive displacement compressors

Positive displacement compressors are those in which successive volumes of air are confined within a closed space and elevated to a higher pressure. The capacity of a positive displacement compressor varies marginally with the working pressure.

Reciprocating compressors

The compressing and displacing element (piston and diaphragm) has a reciprocating motion. The piston compressor is available in lubricated and non-lubricated construction.

Helical and spiral-lobe compressors (screw)

Rotary, positive displacement machines in which two inter-meshing motors, each in helical configuration, displace and compress the air. Available in lubricated and non-lubricated construction. The discharge air is normally free from pulsation; operates at high rotation speed.

Sliding-vane compressors

Rotary, positive displacement machines in which axial vanes slide radially in a rotor which is mounted eccentrically within a cylindrical casing. Available in lubricated and non-lubricated construction; the discharge air is normally free from pulsation.

Two impeller straight-lobe compressors and blowers

Rotary, positive displacement machines in which two straight, mating but non-touching lobed impellers trap the air and carry it from intake to discharge. Non-lubricated. The discharge is normally free from pulsation. Operates at low pressure and high rotation speed.

Dynamic compressors

Dynamic compressors are rotary continuous flow machines in which the rapidly rotating element accelerates the air as it passes through the element, converting the velocity head into pressure, partially in the rotating element and partially in stationary diffusers or blades. The capacity of a dynamic compressor varies considerably with the working pressure.

Centrifugal compressors

Acceleration of the air is obtained through the action of one or more rotating impellers. Non lubricated. The discharge air is free from pulsation. Operates at very high rotation speed.

Axial compressors

Acceleration of the air is obtained through the action of a bladed motor, shrouded at the blade ends. Non-lubricated. Very high rotational speed. High volume output.

Figures 5.1 to 5.7 show diagrammatically the basic operating principles and configuration of some of the more common types of air compressor.

Single-acting, single-stage, vertical, reciprocating compressor

This machine takes in air at atmospheric pressure and compresses it to the required pressure in a single stage. Compression takes place only on the upstroke of the piston.

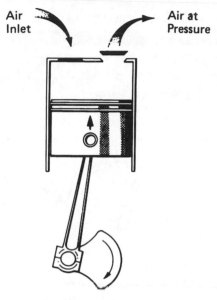

Figure 5.1 Single-acting, single-stage, vertical reciprocating compressor

Single-acting, two-stage, vertical, reciprocating compressor

This machine takes in air at atmospheric pressure and compresses it in two stages to the final pressure. If the final pressure is 7 bar, the first stage normally compresses the air to about 2 bar after which it is cooled at constant pressure. It is then fed into the second-stage cylinder which compresses it to 7 bar. Compression takes place on the upstroke of each piston.

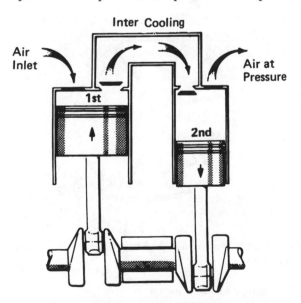

Figure 5.2 Single-acting, two-stage, vertical, reciprocating compressor

Double-acting, two-stage, reciprocating compressor

The difference between Figures 5.2 and 5.3 is that in Figure 5.3 the cylinders are angled and that both sides of each piston are used to compress the air.

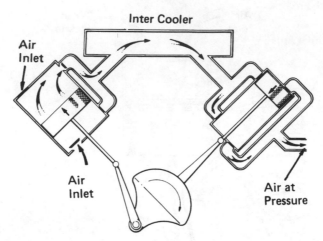

Figure 5.3 V (or angled), double-acting, two-stage, reciprocating compressor

Rotary vane compressor

The rotary vane compressor consists of a rotor mounted eccentrically in a cylindrical chamber. The rotor contains blades free to slide in radial slots. As the rotor rotates so the space between the adjacent blades decreases from air inlet to exhaust. Oil is injected into the machine to act as a seal for the blades against the cylindrical chamber. The oil also acts as a coolant to remove some of the heat generated by compressing the air.

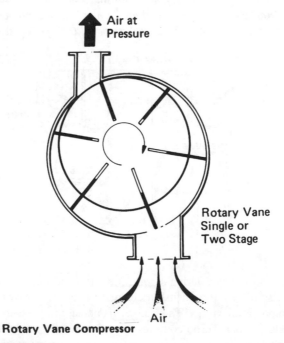

Figure 5.4 Rotary vane compressor

Screw compressor

Two meshing helical rotors rotate in opposite directions. The design of the rotors is such that the free space between them decreases axially in volume and this decrease in volume compresses the air trapped between the rotors.

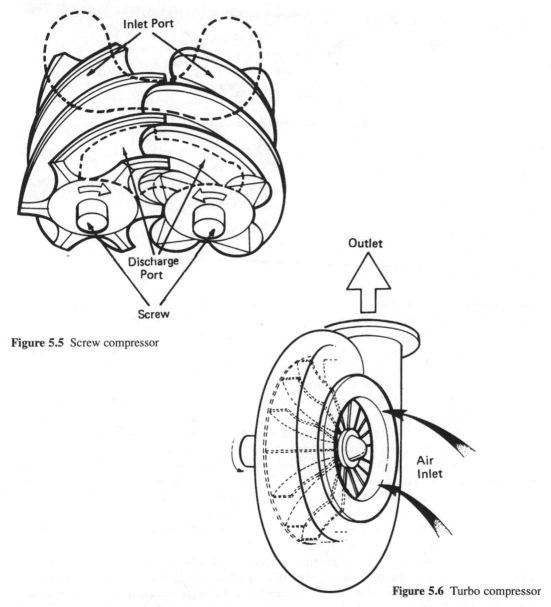

Figure 5.5 Screw compressor

Figure 5.6 Turbo compressor

Turbo compressor

Air is drawn in at the centre of a rotating scroll. The scroll, in rotating imparts kinetic energy to the air. Air leaves the scroll via the diffusion chamber where it is slowed down and the kinetic energy is converted to potential energy. This causes an increase in pressure. Several stages are normally required to produce 7 bar. Turbo compressors are generally used for high air flow rates and can produce oil-free air.

Diaphragm Compressor

These are small single acting units. As the name implies they consists of a diaphragm which rec-
iprocates and by doing so compresses air. They give oil free air and pressures to about 7 bar.

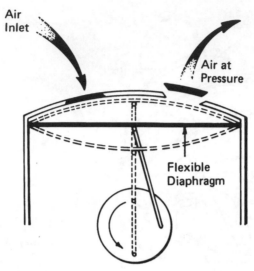

Figure 5.7 Diaphragm compressor

The most common industrial type of air compressor is still the reciprocating machine, although
screw and vane types are finding increasing application, as also are turbo units in the larger sizes.

Some types of compressor can be supplied in 'oil-free' form. In these machines, no oil is
allowed to enter the compression chamber and the following basic modifications are introduced.
In a reciprocating compressor where the piston is sealed with piston rings, the materials are
changed from the normal cast-iron ring which requires oil for lubrication, to either carbon graphite
or PTFE. (PTFE is a self-lubricating material which will withstand the pressures and mechanical
forces involved in the running of the compressor whilst carbon graphite uses the moisture in compressed air as the lubricant.) In a screw compressor the clearances between the mating male and
female rotors are adjusted so that there is no metal-to-metal contact.

The rotary vane (or sliding vane) machine is often described as an 'oil-flooded' unit. Quite large
quantities of oil are injected into the incoming air and pass right through the compression chamber with the air. The oil not only seals the vanes against the periphery of the compression space
but also acts as a lubricant and as a means of removing a fair amount of the heat generated. This
oil is then reclaimed from the compressed air by means of separators, cooled and re-circulated.
This type of machine has the added advantage of being very quiet in operation.

With any lubricated compressor a certain amount of oil is bound to be carried over into the
compressed air system. Provided that air-line filters are used, the amount of such oil is generally
not enough to cause trouble in most industrial applications. In larger quantities, however, this oil
(which retains very little lubricating power) can be a danger in several ways:

- In the initial length of compressed air pipework which is hot, it can be a potential fire hazard.
- It can block small orifices and cause pneumatic valves to stick or operate sluggishly.
- It can cause damage to products if the air is discharged near a manufacturing process.

Air preparation and service units comprising filters, regulators and lubricators are discussed later
in this chapter.

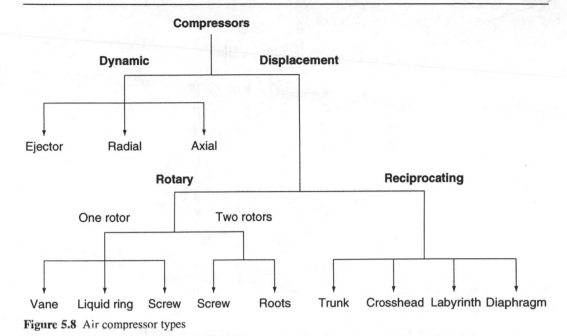

Figure 5.8 Air compressor types

5.3 Coolers and dryers

When air leaves a compressor its temperature is well above that of the surroundings and if we allow it to pass at once through a pipework system it will cool as it goes, depositing water within the pipework and causing rust. What is worse, we shall finish up with water at the point of use. It is good practice, therefore, to cool the air immediately after it leaves the compressor, so that most of this water is condensed out before it reaches the pipework. After-coolers are used for this purpose and can be of either the air-blast or water-cooled type.

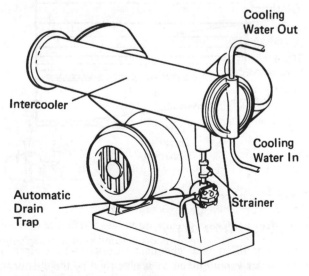

Figure 5.9 Intercooler with automatic drain trap

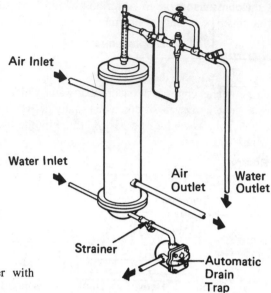

Figure 5.10 Aftercooler with automatic drain trap

Provided they are correctly sized, it is possible to get the temperature of the air leaving them to within 10°C of the water inlet temperature in a water-cooled unit, or to within 15°C of the ambient air temperature if the unit is air-cooled. The water condensed out of the air must of course be drained from the after-cooler, and this is best done through an automatic drain trap. Figures 5.9 and 5.10 show typical arrangements. Some applications need extra dry air and it may be necessary to reduce the dew point by fitting a special air dryer. There are three main types.

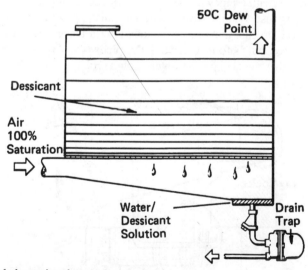

Figure 5.11 Chemical absorption dryer

1 The chemical absorption dryer

This consists of a container filled with chemical through which air is passed as shown in Figure 5.11. Some of the water vapour in the air is absorbed by the chemical, which then turns to liquid and must be drained off.

When all has been used, the containers have to be refilled with a fresh supply of chemical. The dew point of the air leaving this type of unit, at line pressure, is usually about 5°C.

2 The refrigeration dryer

This incorporates a refrigeration unit (like that of a domestic fridge), through which the air passes, as in Figure 5.12. It also contains controls to ensure that water which has been drained out of the air does not freeze inside the unit. Dryers of this type usually deliver air at a line pressure dew point of about 2°C. The cooled air, as can be seen, is re-heated by the incoming air, cooling the latter and increasing the volume of the former.

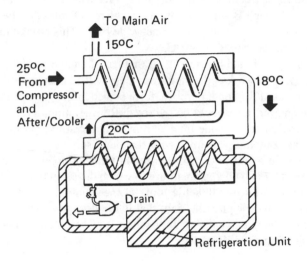

Figure 5.12 Refrigeration dryer

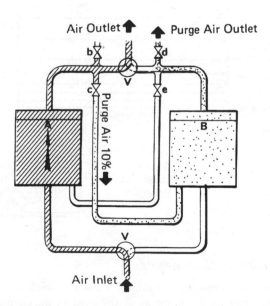

Figure 5.13 Adsorption dryer

3 The adsorption dryer

This consists of twin containers holding a solid chemical with the texture of a fine honeycomb or sponge. This texture gives it a large surface area to which the water droplets will cling. The water is adsorbed onto the surface, leaving the chemical unchanged, instead of being absorbed into the chemical to form a solution as in the chemical absorption dryer. Since the chemical remains unchanged it can be regenerated by heating or by blowing dry air through it. The operation of the dryer is shown in Figure 5.13. With valves **V** in the position shown, the air flows through the vessel **A** where it is dried. Most of it passes to the outflow but, with valves **c** and **d** open, **b** and **e** closed, about 10 per cent of the air is bled off to vessel **B** to re-activate the chemical there.

Before the chemical in **A** becomes saturated, all the valves are switched to their other position, so that the main flow is through **B** while the dry air purge re-activates **A**. The change-over is usually arranged to take place automatically on a time-cycle basis. This type of dryer can bring down the line pressure dew point of the air to −40°C.

Dryers, of whatever type, should be used on air that has already passed through an after-cooler, since otherwise the amount of work demanded from them will be excessive.

When the air is to be used for some purpose, for instance in the food processing industry, it is important that any oil vapour in it should be removed. Refrigeration dryers automatically reduce the oil vapour concentration at the same time as removing moisture, but oil vapour may also be removed in special units containing chemicals for this purpose.

We have described the effectiveness of the different types of dryer by quoting what is known as the 'pressure dew point', that is the temperature to which the air in the pipeline at working pressure can cool before moisture will settle out. Some manufacturers, however, quote the 'atmospheric dew point' for their plant instead, that is the dew point which the air coming from the dryer would have if it were expanded to atmospheric pressure. Caution is needed in interpreting such a figure, as an example will show.

Table 5.1

Dew point °C	Water content of saturated air g/m³
−55	0.021
−50	0.039
−45	0.069
−40	0.12
−35	0.20
−30	0.35
−25	0.55
−20	0.90
−15	1.4
−10	2.1
−5	3.4
0	4.8
5	6.9
10	9.4

Suppose a chemical dryer has a pressure dew point of 5°C at a working pressure of 7 bar. If we compress 1 m³ of air to 7 bar its volume will be reduced to 0.125 m³ (because PV = constant – Chapter 2). From Table 5.1, 1 m³ of air at 5°C can hold 6.9 g of water in vapour form. Therefore 0.125 m³ will be able to hold 0.86 g. If we allow the 0.125 m³ of air to expand back to its original volume, 1 m³, we now have at atmospheric pressure 1 m³ of air containing 0.86 g of water in vapour form. From Table 5.1, we can see that this water can be contained in the air at a temperature of a little below –20°C (because at this temperature 1 m³ of air can contain 0.9 g of water) but not as low as –25°C (because at that temperature the air can only hold 0.55 g of water vapour). By interpolation we find that at –20.5°C 1 m³ of air can just about hold 0.86 g of water vapour.

So what does this tell us? First, that a pressure dew point of 5°C at 7 bar is the same as an atmospheric dew point of –20.5°C. But a more important point is that, if the manufacturer has quoted the atmospheric dew point and the air is to be used out of doors in winter conditions, we might be tempted to suppose that there is no danger of instrument freezing unless the temperature drops to –20.5°C since above that temperature the water will remain in vapour form and cannot freeze. The catch is that this argument only applies to the air after it has been discharged. In the pipeline and in the pneumatic equipment, water will condense out as soon as the temperature of the compressed air falls below 5°C, and ice will be formed as soon as it falls to 0°C.

You may have noticed that in the calculation giving the atmospheric dew point no allowance was made for the change in volume of the air when it is cooled from 5°C. This change would have had little effect, since the absolute temperature drops from 278 K to 252.5 K, a decrease of 9 per cent, while the water content drops from 6.9 to 0.86, a decrease of nearly 90 per cent. If we had allowed for change of volume we should have found the atmospheric dew point to be –20°C instead of –20.5°C a difference of no practical importance.

5.4 The air receiver

Receivers provide constant air pressure in a pneumatic system, regardless of varying or fluctuating consumption (Figure 5.14). This enables briefly occurring consumption peaks to be balanced out, which cannot be made up by the compressor.

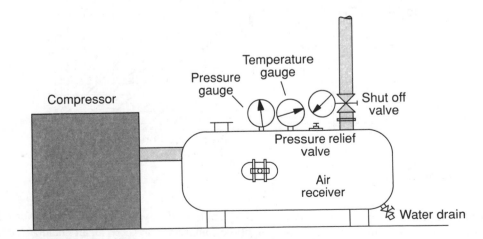

Figure 5.14 Air receiver installation

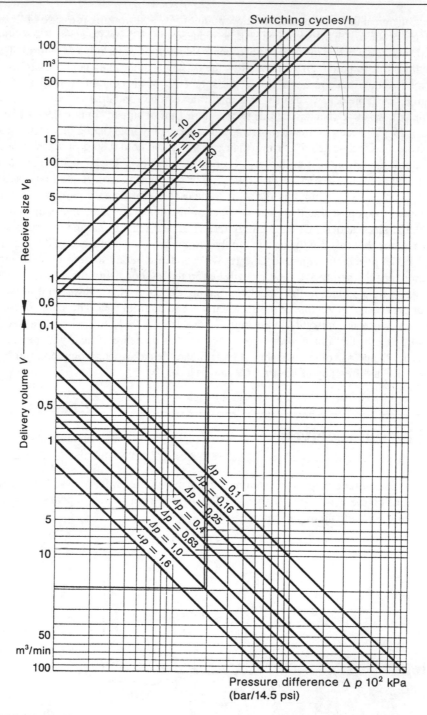

Figure 5.15 Air receiver size chart

A further function of receivers is the emergency supply to the system in cases of power failure. The reservoir can be fitted either downstream of the compressor, to act as an air chamber, or selectively at points where consumption is high.

In addition, the large surface area of the receiver cools the air, which is particularly advantageous in systems which do not have chemical or freeze dryers fitted. Thus, a portion of the moisture in the air is separated directly from the receiver as water. It is therefore important to drain the condensate regularly.

The size of a compressed air receiver depends upon the:

- delivery volume of the compressor
- air consumption for the applications
- network size
- type of compressor cycle regulation
- permissible pressure drop in the supply network.

Air receiver size

Example

Delivery volume	$V = 20 \text{ m}^3/\text{min}$
Pressure drop	$\Delta P = 1 \text{ bar}$
Switching cycles per hour	$z = 20$

Result:

| Receiver size | $V_B = 15 \text{ m}^3$ |

(refer to Figure 5.15 on the previous page)

Figures 5.16 and 5.17 illustrate a vertically mounted receiver and it will be noticed that automatic drain traps should be fitted to all receivers so as to ensure that moisture is not allowed to build up inside them.

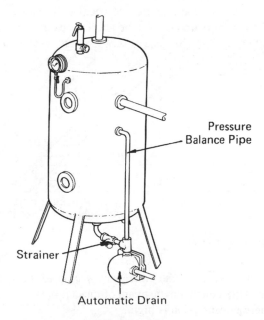

Figure 5.16 Vertically mounted receiver with drain trap

In some arrangements sludge may also form inside the receiver when the compressor becomes worn. This sludge may well be too thick for the automatic drain trap to handle and on such systems the trap may be fitted as shown in Figure 5.17 whereby a manually operated tap is connected into the vessel at a higher level. The scum and oil can therefore be drained off by an operator at either weekly or monthly intervals and therefore before it is allowed to reach the automatic drain trap.

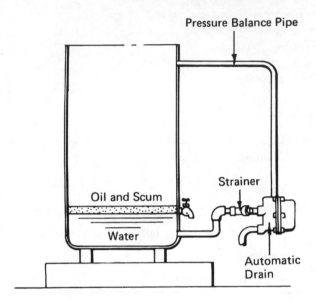

Figure 5.17 Section through vertically mounted receiver showing manual drain off facility for oil scum

5.5 Automatic drain traps

As we have seen, moisture condenses out not only when air is compressed but also when it is cooled. In both instances we have to provide adequate means for draining the water off. In this respect automatic drain traps are used. There are devices that will open and allow any accumulated water to be discharged, but will close as soon as all the water has been expelled thus preventing air from escaping.

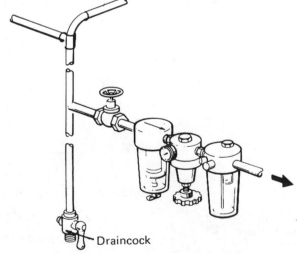

Figure 5.18 Manual drain cock facility on air service pipework

The use of manual drain cocks, as shown in Figure 5.18, is not very satisfactory because it relies on people to open them at regular intervals. If they are forgotten they can cause all the liquids which have been separated out of the air to be re-entrained and carried over into the system with possible breakdown of pneumatic equipment and spoilage of product, both of which can be very expensive. It makes sense, therefore, to use some sort of automatic device which will release any liquid but remain closed to air. There are a number available on the market, each having its own particular characteristics as follows.

Float-operated traps can employ either a float directly in contact with an orifice, as shown in Figure 5.19, or a float which opens the valve through a lever arm, as shown in Figure 5.20. The advantage of the float-operated type is that it is simple and positive. The disadvantage is that, because of the mechanical force needed to open the valve, the discharge orifice has to be small, particularly the directly operated float type, and it can easily get blocked by the dirt, silt or gum which are so often found in compressed air systems. This type of trap may also be unable to deal with foam which sometimes comes over from compressors and may not be dense enough to lift the float.

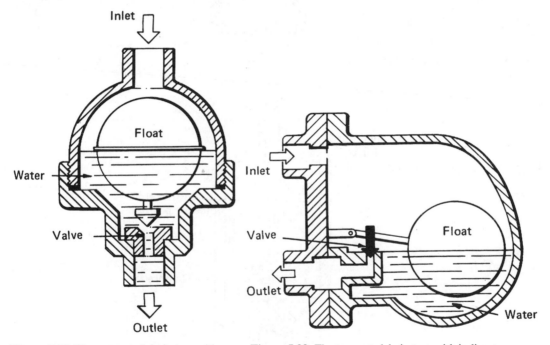

Figure 5.19 Float operated drain trap with float directly in contact with orifice

Figure 5.20 Float operated drain trap with indirect operation of valve via lever arm

An alternative type is one having an inverted bucket as a float (see Figure 5.21). When the bucket is surrounded by water, air trapped inside it bleeds out very slowly through a small hole in the top, causing the bucket to lose buoyancy and fall. In so doing it opens a valve through which the water is discharged by the line pressure. So long as accumulated water feeds into the trap the valve remains open, but when air enters it displaces the water in the bucket, with the result that the bucket gains buoyancy, rises, and closes the valve. The cycle is then repeated.

This type of unit can handle foam but it does need initial priming with water. It has an intermittent blast action that can be useful when conditions are particularly dirty.

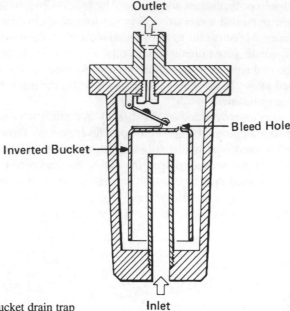

Figure 5.21 Inverted bucket drain trap

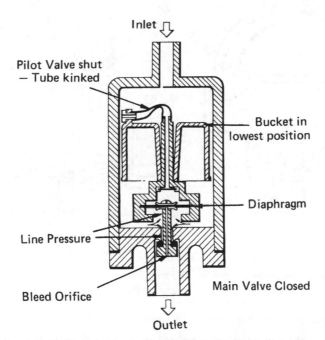

Figure 5.22 Alternative design of inverted bucket drain trap

The third type of unit, shown in Figure 5.22 again uses an inverted bucket float but is pilot-operated. The pilot valve is a rubber tube which kinks when the float is in the low position, preventing air from flowing through it to the upper side of a diaphragm connected to the discharge valve. The area of the diaphragm is greater than that of the valve, with the result that, so long as the pressure above the diaphragm remains atmospheric, the line pressure holds it up and keeps the valve closed. As water accumulates in the unit however, the float rises, opens the pilot valve, and allows the pressure on both sides of the diaphragm to equalise. The main valve will then be forced open by the system pressure and the liquid in the trap will be discharged. As soon as all the water is out of the system the float falls and the rubber tube kinks again as shown. The air above the diaphragm bleeds away through a small bleed hole down the length of the valve stem, and finally the difference in pressure on the two sides of the diaphragm causes the discharge valve to close. This type of unit, being pilot operated, can use a much larger discharge orifice than the previous type and will cope better with contaminated liquid.

Another type uses, as its operating principle, the fact that if you bring a fast-moving fluid to rest there will be a build-up of pressure. This type is shown in Figure 5.23 and consists essentially of a shallow cylindrical container whose flat base has a circular inlet port in the middle with a concentric circular groove cut round it, leaving two flat concentric rings. One or more outlet ports lead out from the bottom of the groove. A flat disc is the only moving component.

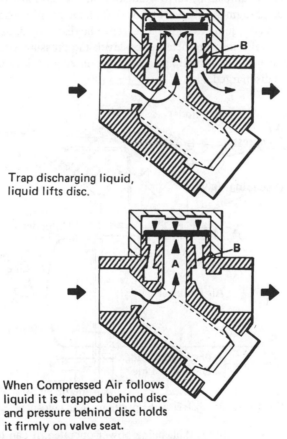

Trap discharging liquid, liquid lifts disc.

When Compressed Air follows liquid it is trapped behind disc and pressure behind disc holds it firmly on valve seat.

Figure 5.23 Operation of a drain trap; applying the Bernoulli theorem

When water flows up through the central port A it will push the disc back towards the cap and flow across the underside of the disc through to the outlet port B. When, however, a compressible fluid like air follows the water, it expands radially across the underside of the disc and so will travel at a very much higher velocity. This means that the static pressure under the disc is reduced, making use of the Bernoulli theorem which states that the sum of the pressure and velocity energy in a flowing fluid remains constant. Owing to this lowering of pressure the disc starts to move down towards the seating lands, some of the air is trapped in the cap behind the disc and in being trapped comes to rest. Again by the Bernoulli theorem there is a build-up in pressure behind the disc and the disc will snap shut and remain closed as long as the force due to the pressure in the cap acting over the outer seating land is greater than the force from the compressed air system acting over the inlet port area. The pressure in the cap gradually falls because the trap is made in such a way as to allow a very small constant bleed of air, and when it has fallen sufficiently the disc will move away from the inlet port and will again either allow water to be discharged or, if there be none, snap shut immediately due to the air flow (as detailed above).

There is yet another type of drain. It works off the compressor unloading valve, a pressure switch or an electric timer coupled to a solenoid-operated three-way valve. Refer to Figure 5.24. To understand its operation, look at Figure 5.25. In **A** of this diagram there is zero pressure in the connection from the compressor unloading valve (that is, the receiver is not yet up to the cut-out pressure of the compressor) and the valve is closed. In **B** the receiver is up to the unloading pressure and the signal from the unloading valve which unloads the compressor also actuates the drain valve through a large diaphragm. As the valve moves from the position shown in **A** to the position in **C**, it allows accumulated liquids from the receiver to be discharged, as shown in **B**. When the pressure in the receiver falls to a pre-determined level, the pressure signal from the unloading valve is removed and then the drain will again open (in moving back to position **A**) as shown in **D**. Once again liquid is discharged.

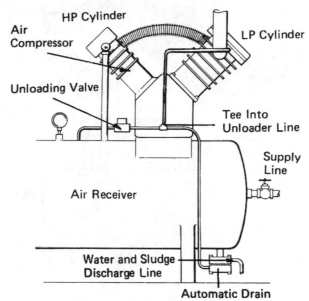

Figure 5.24 Drain trap fitted to an air receiver with operation via an unloading valve

An advantage of this type of valve is that, being power-operated, it can have a large bore valve seat and it is good, therefore, at handling viscous sludge sometimes found in receivers – particularly if the compressor is in need of overhaul. It has, however, the disadvantage of operating, not when moisture has accumulated, but either on a time-cycle or every time the compressor unloads. It can, therefore, either leave accumulated liquids in the system or waste air, depending on the setting; and it can never operate at exactly the right rate for all conditions. It has the further disadvantage of being very much more expensive than any other types shown.

With such a large choice of automatic drains available, we must ask ourselves which one we should choose. There is no hard and fast rule here; much obviously depends on individual preferences. However, here are some suggestions.

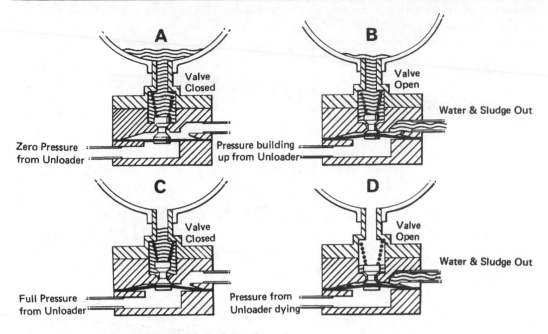

Figure 5.25 Operation of draintrap via an unloading valve

Provided the condensate is not too emulsified and sticky, the float-operated traps are the simplest and most positive. If neatness is of importance or you have any doubts about the straightforward float-operated type, choose one of the float pilot traps. If there is a lot of 'foam' or dirt in the system, the inverted bucket can sometimes cope where the float can't (but don't forget to prime this type when first fitted). In particularly dirty conditions – and for use out of doors when freezing can be a hazard – the disc type can be very useful. The one working off the compressor unloading valve or timer can also cope well with dirt and sludge, but may be damaged by frost and is not nearly as simple to install.

Before finally leaving automatic drains there is one other aspect which we must consider – how to fit them. Provided the inlet to the drain is comparatively large and it is fitted close to the equipment it is draining (a matter of a few centimetres away) there is no problem. However, where a substantial amount of liquid can form quickly or if the liquid is likely to be viscous, particularly when the drain is connected to the plant it is draining by a comparatively long length of horizontal pipe, a phenomenon known as 'air-binding' may occur. This is shown in Figure 5.26, which illustrates that unless the liquid can get into the drain (and it can only do this, of course, by displacing air out of it), it is impossible for the liquid to be discharged. For this reason many automatic traps are provided with a means of connecting a balance pipe, and if they are fitted some distance away, as shown in Figure 5.27, the balance pipe should be connected back into the vessel being drained. Under these conditions, any liquid flowing into the drain can easily displace air already in the drain. If by any chance there is no provision for a balance pipe, a T-connection should be fitted to the inlet of the trap.

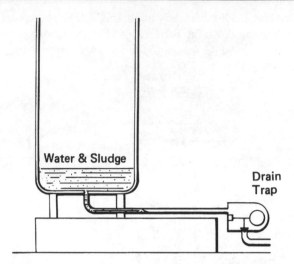

Figure 5.26 Effect of air binding

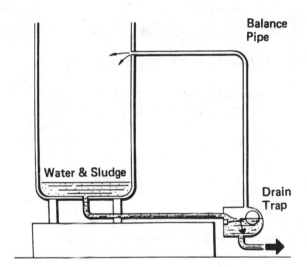

Figure 5.27 Installation of balance pipe to mitigate the effect of air binding

5.6 The distribution system

In ideal circumstances of cooling and separation all oil and water should have settled out before the air leaves the receiver. In practice, this very seldom happens unless expensive dryers are fitted. The compressed air distribution pipework therefore acts as an additional cooling surface and water and oil will separate out throughout its length. To assist drainage, the pipework should be given a fall of about 1–2 per cent in the direction of flow and it should be adequately drained.

It is useful to lay out a distribution system in the form of a ring main if possible. This will help to lower the velocity of air in the main, preventing it from re-entraining deposited moisture and reducing the pressure drop. Moreover, it will mean that near the point of use there is a store of compressed air which will help if a sudden draw-off occurs. If at some future date, extra air users are installed and another branch is fitted, this will be able to draw some of its air from both directions, thus keeping down the increase in velocity. Drainage points should be provided by using equal tees. It assists in the separation of the water if these are arranged to change the direction of flow, as shown in Figure 5.28.

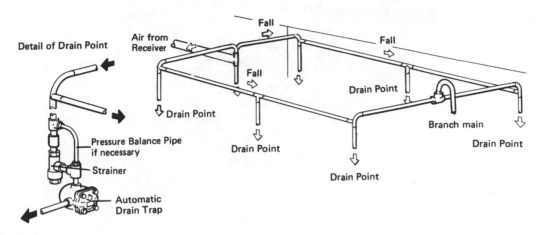

Figure 5.28 Air distribution system showing drainage points

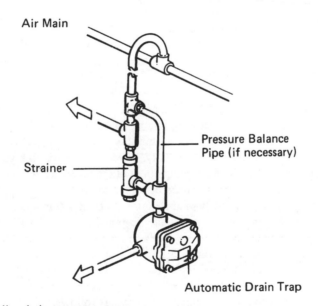

Figure 5.29 Branch line drainage arrangement

Any branch line should be taken off the top of the main, to prevent water in the main pipe from running into it, and the bottom of the falling pipe should be drained as in Figure 5.29.

Although the pipe layout suggested above will effectively deal with any water which has collected at the bottom of the main, it can do little to separate out the mist of water droplets which are sometimes suspended in air, particularly when velocity surges occur. An excellent way of dealing with this problem is to fit a separator in the main or sub-main. The principle of a separator is to separate out water droplets either by making them impinge on metal plates which are effectively in their path (the air is made to change direction rapidly over a short length) or by centrifugal means. A separator illustrating the first principle is shown in Figure 5.30. Naturally, the separator requires drainage and an automatic drain trap should be fitted.

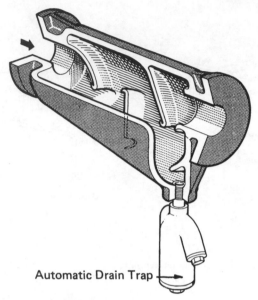

Automatic Drain Trap

Figure 5.30 Automatic drain trap fitted to separator

Sizing

The next point to consider is the sizing of the mains. Although large mains are more expensive than small ones it is thoroughly bad practice to undersize them. In the first place undersizing can lead to excessive pressure drop along the mains, and this can be serious if it prevents equipment from operating as efficiently as it should. Secondly, undersizing leads to high air velocities, and any water that condenses out through the cooling effect of the pipework will be whipped up by the fast-moving air and carried along with it instead of separating out.

Air mains are usually sized on velocity, and a figure of about 6–9 m/s is common because this is sufficiently low to prevent an excessive pressure drop and should also allow reasonable water separation. If the main is a new one it usually pays to make an allowance for future increase in demand. In the case of sub-circuits working at pressures of about 6 bar, however, it is perfectly acceptable, provided such circuits are no more than a few metres of pipe, to work at velocities up to about 20 m/s.

A simple example on how to size on velocity is shown as follows:

To find the pipe size to carry an air flow rate of 12 litre/s at a velocity of 9 m/s we calculate as follows:

$$V = \frac{Q \times 1000}{d^2}$$

where:

V = velocity in m/s; Q = air flow rate in the system in litre/s; d = internal pipe diameter in mm.

Therefore:

$$9 = \frac{12 \times 1000}{d^2}$$

$$d = \sqrt{\frac{12 \times 1000}{9}}$$

= 36.5 mm ID.

Referring to Table 5.2, it can be seen that the nearest size pipe up is 40 mm medium weight steel tube (to BS 1387) which has a minimum internal diameter of 41.5 mm and this will satisfy the condition.

Table 5.2 Some standard tube dimensions (Steel tubes BS 1387)

Nominal bore (mm)	Med weight min ID (mm)	Heavy weight min ID (mm)
6.0	5.8	4.5
8.0	8.6	7.5
10.0	12.1	11.0
15.0	15.8	14.6
20.0	21.3	20.1
25.0	26.9	25.3
32.0	35.6	34.0
40.0	41.5	39.9
50.0	52.5	50.8
65.0	68.1	66.4
80.0	80.0	78.4
100.0	104.0	102.0
125.0	129.0	128.0
150.0	154.0	153.0

Pressure drop

When sizing both mains and sub-circuits it is as well to check back on the pressure drop to ensure that it will not be excessive. To do this it is first necessary to find the equivalent pipe run, that is the true length in metres plus an allowance for bends and fittings. Table 5.3 shows the pressure loss through steel pipe fittings as equivalent pipe lengths in metres. Often a nomogram is used to ascertain the pressure drop through pipes, as illustrated in Figure 5.31.

It should be noted that for copper or ABS pipes, the pressure drop is about 20 per cent less than for a steel pipe of the same internal diameter.

Table 5.3 Resistance of pipe fittings (equivalent length in m)

Type of fitting	Nominal pipe size									
	15	20	25	32	40	50	65	80	100	125
Elbow	0.26	0.37	0.49	0.67	0.76	1.07	1.37	1.83	2.44	3.20
90° bend (long)	0.15	0.18	0.24	0.38	0.46	0.61	0.76	0.91	1.20	1.52
Return bend	0.46	0.61	0.76	1.07	1.20	1.68	1.98	2.60	3.66	4.88
Globe valve	0.76	1.07	1.37	1.98	2.44	3.36	3.96	5.18	7.32	9.45
Gate valve	0.107	0.14	0.18	0.27	0.32	0.40	0.49	0.64	0.91	1.20
Run of standard tee	0.12	0.18	0.24	0.38	0.40	0.52	0.67	0.85	1.20	1.52
Through side outlet of tee	0.52	0.70	0.91	1.37	1.58	2.14	2.74	3.66	4.88	6.40

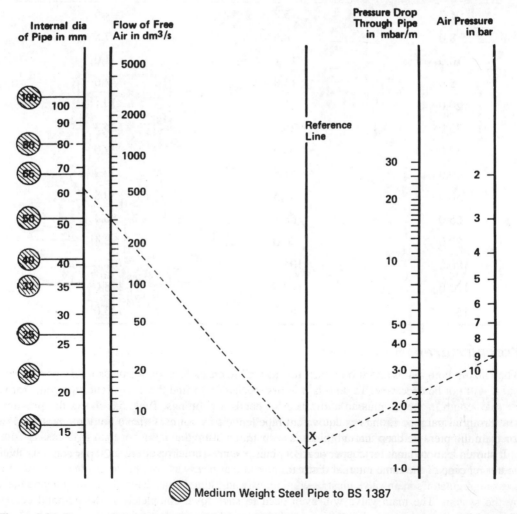

Figure 5.31 Pressure drop in steel pipes (15 mm to 100 mm)

5.7 Air preparation

It is essential to have good quality air in order to promote reliable and effective operation of pneumatic components. To this end it is usual to prepare the air after generation with an air service unit comprising a filter, regulator and lubricator. This is also sometimes referred to as a FRL unit.

Filters

The task of a filter is to separate and collect water droplets and solid particles that are present in the air flowing through the filter. Unless the air has previously been processed through air-drying equipment it will be flowing with a relative humidity of 100 per cent. This 100 per cent value is the maximum amount of water that it is possible for air to hold in a vapour form and this will vary with the compressed air temperature. The lower the temperature the less the amount of water vapour.

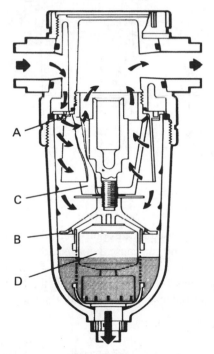

Figure 5.32 Operation of a typical filter

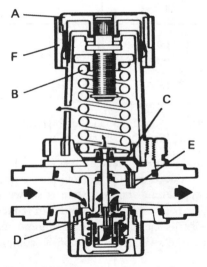

Figure 5.33 Operation of a typical pressure regulator

Regulators

The essential purpose of the regulator is to reduce the primary pressure to the most suited value for the equipment and application that it is serving. Pressure that is higher than is necessary simply wastes energy and increases wear on components, both resulting in unnecessarily high operating costs. To maintain the downstream pressure as nearly as possible to a constant value, the pressure regulator's main valve must be continuously adjusting to match the changing demand made by the system. The main valve will even need to shut down completely if the demand ceases. Figure 5.33 shows a section through a typical modern relieving-type regulator.

Lubricators

It is usual for sliding parts to require lubrication in order to reduce wear and frictional losses. Sliding parts are contained in pneumatic devices (cylinders and valves etc.). To ensure that they are continually supplied with an adequate amount of lubricant, a certain quantity of specially selected oil is added to the compressed air by means of a lubricator. The compressed air then supplies the oil particles to the elements. Diagrams of typical oil mist lubricators are shown in Figure 5.34.

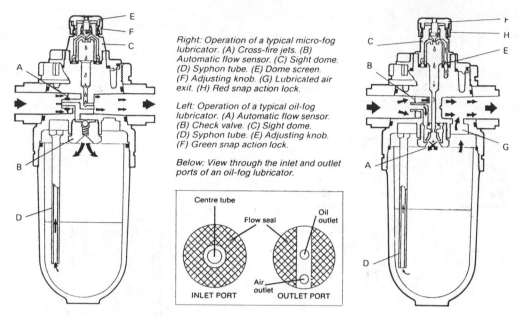

Right: Operation of a typical micro-fog lubricator. (A) Cross-fire jets. (B) Automatic flow sensor. (C) Sight dome. (D) Dome screen. (E) Adjusting knob. (F) Adjusting knob. (G) Lubricated air exit. (H) Red snap action lock.

Left: Operation of a typical oil-fog lubricator. (A) Automatic flow sensor. (B) Check valve. (C) Sight dome. (D) Syphon tube. (E) Adjusting knob. (F) Green snap action lock.

Below: View through the inlet and outlet ports of an oil-fog lubricator.

Figure 5.34 Right: Operation of a typical micro-fog lubricator. Left: Operation of a typical oil-fog lubricator. Centre: View through the inlet and outlet ports of an oil-fog lubricator.

It should be noted that only specially selected lubricants are to be used for this purpose. Oils which are introduced into the air from the compressor are not suitable and can cause damage and failure.

A typical air supply system is shown in Figure 5.35. As previously mentioned in **5.4**, an air receiver is usually fitted in order to reduce pressure fluctuations. In normal operations, the compressor fills the receiver when required and the receiver is available as a reserve at all times. This reduces the switching cycles of the compressor.

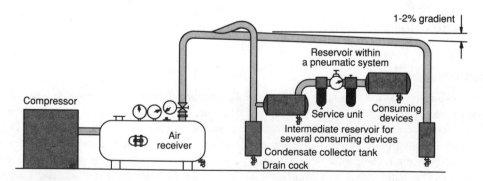

Figure 5.35 Air supply system

Poorly prepared compressed air will inevitably lead to malfunctions and may manifest itself in the system as follows:

- rapid wear of seals and moving parts in the cylinders and valves
- oiled-up valves
- contaminated silencers.

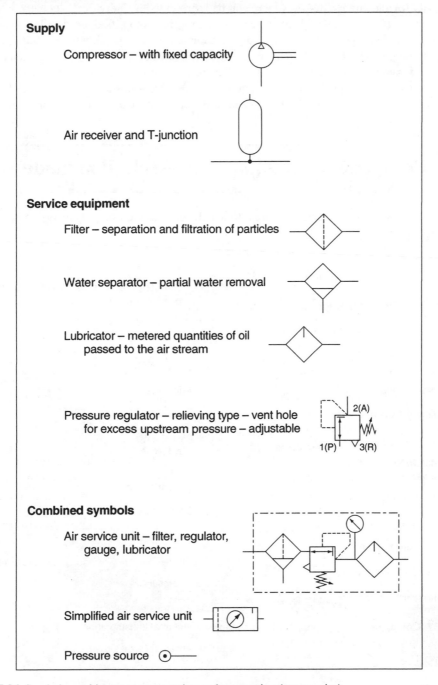

Figure 5.36 Symbols used in energy conversion and preparation (pneumatics)

5.8 Graphical symbols used in energy conversion and air preparation

The symbols for the energy supply system can be represented as individual elements or as combined elements (Figure 5.36). The choice between using simplified or detailed symbols is dependent upon the purpose of the circuit and its complexity. In general, where specific technical details are to be given such as requirements for non-lubricated air or micro-filtering, then the complete detailed symbol should be used. If a standard and common air supply is used for all components, then the simplified symbols can be used. For trouble-shooting the detailed symbols are more suitable. But the detail should not add to the complexity of the circuit for reading.

5.9 Compressed air equipment – selection guide

Figures 5.37, 5.38 and 5.39 indicate the areas to which consideration should be given when designing a compressed air system.

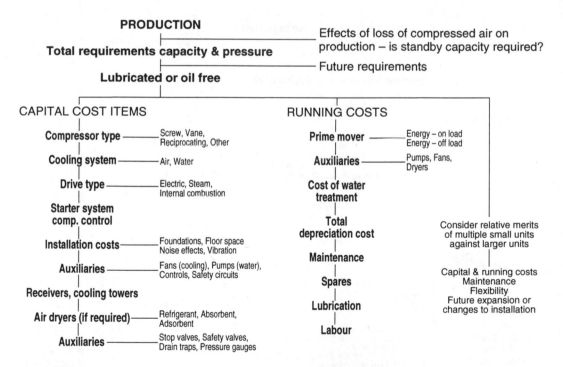

Figure 5.37 Compressed air equipment selection guide, commercial considerations

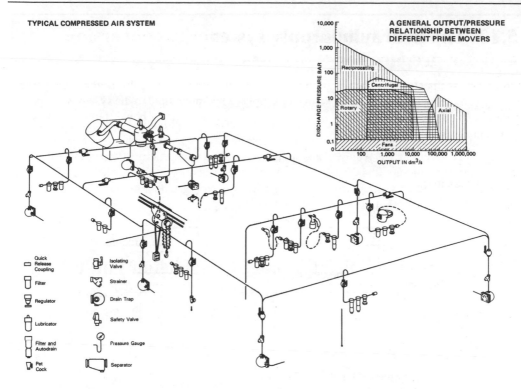

Figure 5.38 Typical compressed air system: a general output/pressure relationship between prime movers

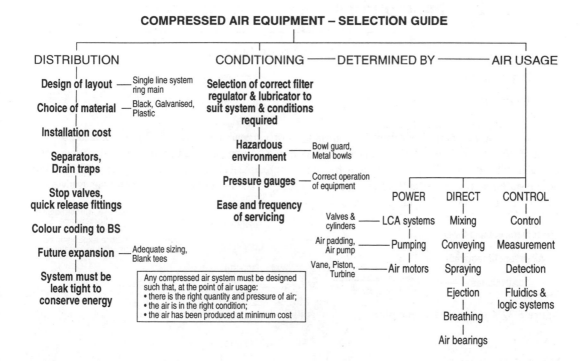

Figure 5.39 Compressed air equipment selection guide, technical and engineering considerations

5.10 The hydraulic supply system

A simple arrangement of a hydraulic supply system is shown in Figure 5.40. In this arrangement a primary storage tank is the principal source of supply of hydraulic fluid to the system via a pump. Accumulators can be used to 'top-up' or act as a reserve in the event of power failure or if excessive demand is made upon the system.

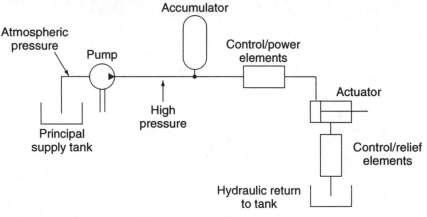

Figure 5.40 Typical arrangement for a hydraulic supply system

5.11 Hydraulic pumps

Three types of hydraulic pump are generally in use:

- gear pump
- vane pump
- piston pump.

The most popular is the gear pump, comprising a drive gear that is rotated by an external source via a shaft. This is meshed with a driven gear within a housing consisting of inlet and outlet ports. Fluid is forced through the pump as it becomes trapped between the rotating gear teeth and the housing. Thus low pressure fluid enters the chamber via the inlet port and is expelled as high pressure fluid at the outlet. Figure 5.41 shows the arrangement.

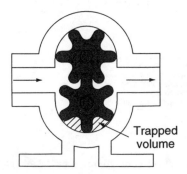

Figure 5.41 Hydraulic external gear pump

5.12 Accumulators

Two basic types of accumulator are in use:

- gas-pressurised accumulator
- spring-loaded accumulator.

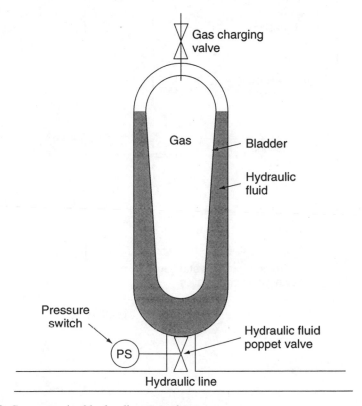

Figure 5.42 Gas pressurised hydraulic accumulator

The majority of accumulators are of the gas-pressurised type whereby gas is pressurised within a bladder with hydraulic fluid contained in the chamber between the bladder and the wall of the vessel. Figure 5.42 shows an arrangement. Smaller and older type accumulators may be of the spring-loaded design.

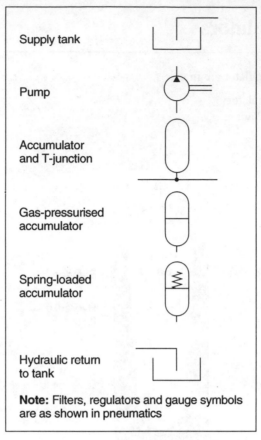

Figure 5.43 Symbols used in energy conversion and preparation (hydraulics)

6 Control Valves I – Types and Principles of Operation

Aims

At the end of this chapter you should be able to:

1 Have an awareness of the various types of valves used in pneumatics and hydraulics.
2 Appreciate the features and operation of the directional control valve in its role as a signalling, processing and power element.
3 Appreciate the features and operation of the non-return valve and its derivatives.
4 Appreciate the features and operation of the flow control valve.

6.1 Types of valves

Valves can be divided into a number of groups according to their function in relation to signal type, actuation method and construction. The primary function of the valve is to alert, generate or cancel signals for the purpose of sensing, processing and controlling. Additionally, the valve is used as a power element for the supply of working fluid to the actuator. Valves may be sub-divided into the following categories.

Dealt with in this chapter:

- directional control valves
  - signalling elements
  - processing elements
  - power elements
- non-return valves
- flow control valves.

Dealt with in Chapter 7:

- pressure control valves
- combinational valves
- solenoid valves.

6.2 Directional control valves – general

The directional control valve controls the passage of fluid signals by generating, cancelling or redirecting signals.

In the field of control technology, the size and construction of the valve is of less importance than the signal generation and actuation method. Directional control valves can be of the poppet or slide type, with the poppet used for small flow rates and for the generation of input and process signals. The slide valve is able to carry larger flow rates and hence lends itself to use in power and actuator control.

The multi-way directional control valve is described by:

- number of ports or openings (ways): 2-way, 3-way, 4-way, 5-way, etc.
- number of positions: 2 positions, 3 positions, etc.
- method of actuation of the valve: manual, air pilot, solenoid, etc.
- methods of return actuation: spring return, air return, etc.
- special features of operation: manual overrides, etc.

As described in Chapter 4, the number of controlled connections and the number of positions on a directional control valve can be represented by either a numbering or lettering system. This is reproduced here to remind you (Figure 6.1).

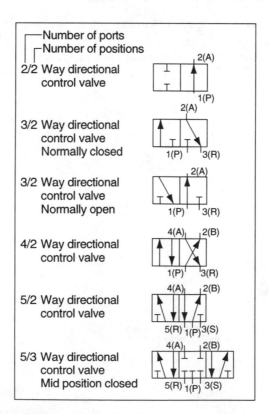

Figure 6.1 Directional control valves: ports and positions (ways)

Port or connection	ISO 5599 numbering system	Lettering system
Pressure port	1	P
Exhaust port	3	R (3/2 way valve)
Exhaust port	5, 3	R, S (5/2 way valve)
Signal outputs	2, 4	B, A
Pilot line opens flow 1 to 2	12	Z (single pilot 3/2 way valve)
Pilot line opens flow 1 to 2	12	Y (5/2 way valve)
Pilot line opens flow 1 to 4	14	Z (5/2 way valve)
Pilot line flow closed	10	Z, Y
Auxiliary pilot air	81, 91	Pz

6.3 The directional control valve as a signalling element

As a signal element the directional control valve is operated by, for example, a roller lever to detect the piston rod position of a cylinder. The signal element can be smaller in size and create a small fluid pulse. A signal pulse created will be at full operating pressure but have a small flow rate.

Figure 6.2 3/2-way roller lever valve

The operation of the 3/2-way roller lever valve in sensing the operation of a cylinder is shown in Figure 6.3.

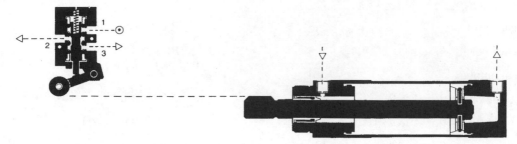

Figure 6.3 Operation of a 3/2 roller lever valve in sensing the displacement of a cylinder

Port 1 on the valve is normally closed until the cam on the cylinder piston rod operates the roller mechanism. When this happens, port 1 is switched to the output port 2 and a signal is given to prove the cylinder has completed its stroke. When the piston rod returns, the valve regains the normal state, putting the output to exhaust through port 3.

6.4 The directional control valve as a processing element

As a processing element the directional control valve redirects, generates or cancels signals depending on the signal inputs received. The processing element can be supplemented with additional elements, such as the AND-function and OR-function valves to create the desired control conditions. This is further discussed in Chapter 9.

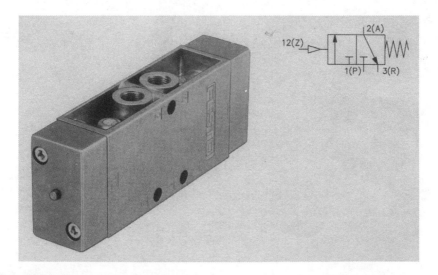

Figure 6.4 3/2-way air actuated valve; single pilot valve

6.5 The directional control valve as a power element

As a power element the directional control valve must deliver the required quantity of fluid to match the actuator requirements and hence there is a need for larger volume flow rates and therefore larger sizes. This may result in a larger supply port or manifold being used to deliver the fluid to the actuator.

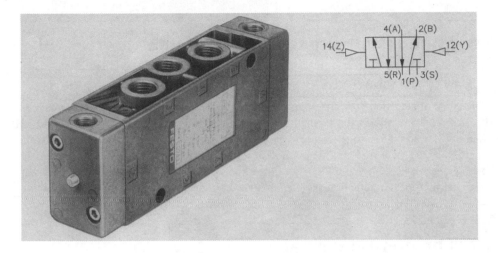

Figure 6.5 5/2-way air actuated valve; double pilot valve

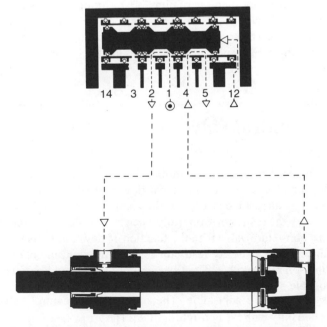

Figure 6.6 Example of a 5/2 directional control valve controlling a double acting cylinder

5/2-way valve for cylinder control: double pilot valve

An example of a 5/2-way directional control valve controlling a double-acting cylinder is shown in Figure 6.6, which is on the previous page. This illustrates a double pilot pressure operated spool valve. The air supply is connected to port 1 which is switched alternately to ports 2 and 4 each time the spool is moved from side to side. A pulse in signal port 12 has moved the spool to switch port 1 to port 2 causing the cylinder to be instroked. A pulse in signal port 14 will move the spool to the right to switch port 1 to port 4 causing the cylinder to outstroke. In the 12 state, port 4 will exhaust to port 5 and in the 14 state, port 2 will exhaust to port 3.

6.6 Methods of actuation – directional control valves

The methods of actuation of directional control valves are dependent upon the requirements of the task. The types of actuation vary, for example mechanical, pneumatic, hydraulic, electrical and combined actuation. The symbols for the methods of actuation (Figure 6.7) are detailed in ISO 1219.

When applied to a directional control valve, consideration must be given to the method of initial actuation of the valve and also the method of return actuation. Normally these are two separate methods. They are both shown on the symbols either side of the position boxes. There may also be additional methods of actuation such as manual overrides, which are indicated separately.

6.7 The non-return valve

The non-return valve allows a signal to flow through the device in one direction and in the other direction blocks the flow. There are many variations in construction and size derived from the basic non-return valve. Other derived valves utilise features of the non-return valve by the incorporation of non-return elements. The non-return valve can be found as an element of the one-way flow control valve, quick exhaust valve, shuttle valve and the two-pressure valve. Refer to symbols in Figures 6.8 and 6.10.

6.8 The flow control valve

The flow control valve (Figure 6.9) restricts or throttles the fluid in a particular direction to reduce the flow rate of the fluid and hence control the signal flow. If the flow control valve is left wide open then the flow should be almost the same as if the restrictor were not fitted. In some cases it is possible to infinitely vary the restrictor from fully open to completely closed. If the flow control valve is fitted with a non-return valve then the function of flow control is uni-directional with full free flow in one direction. A two-way restrictor restricts the fluid in both directions of flow and is not fitted with the non-return valve. The flow control valve should be fitted as close to the working element as possible and must be adjusted to match the requirements of the application.

Most flow control valves are adjustable. The one-way flow control valve permits flow adjustment in one direction only with the non-return fitted. The arrow (see Figure 6.10) shows that the component is adjustable but does not refer to the direction of flow; it is diagrammatic only.

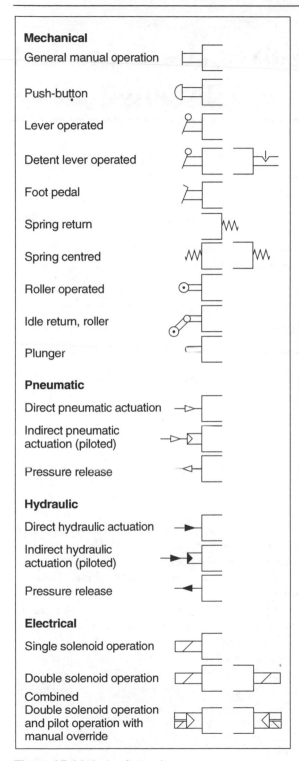

Mechanical

General manual operation

Push-button

Lever operated

Detent lever operated

Foot pedal

Spring return

Spring centred

Roller operated

Idle return, roller

Plunger

Pneumatic

Direct pneumatic actuation

Indirect pneumatic actuation (piloted)

Pressure release

Hydraulic

Direct hydraulic actuation

Indirect hydraulic actuation (piloted)

Pressure release

Electrical

Single solenoid operation

Double solenoid operation

Combined Double solenoid operation and pilot operation with manual override

Figure 6.7 Methods of actuation

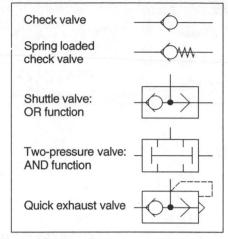

Check valve

Spring loaded check valve

Shuttle valve: OR function

Two-pressure valve: AND function

Quick exhaust valve

Figure 6.8 Non-return valves and derivatives

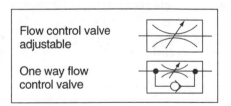

Figure 6.9 Flow control valve

Flow control valve adjustable

One way flow control valve

Figure 6.10 Flow control valves – symbols

7   Control Valves II – Types and Principles of Operation

Aims

At the end of this chapter you should be able to:

1 Appreciate the features and operation of pressure control valves in their role as regulation, relief and sequencing devices.
2 Appreciate the features and operation of combinational valves as time delay devices.
3 Appreciate the features and operation of solenoid valves.

7.1 Pressure control valves

Pressure control valves are used in fluid power systems. There are three main groups:

- pressure regulating valves
- pressure limiting valves
- pressure sequence valves.

The pressure regulating valve controls the operating pressure in a control circuit and keeps the pressure constant irrespective of any pressure fluctuations in the system.

Pressure regulating valve as used in pneumatic circuits

In Figure 7.1, the pressure is regulated by a diaphragm. The output pressure acts on one side of the diaphragm and the spring acts on the other side. The spring force can be adjusted by means of an adjusting screw. When the output pressure increases, the diaphragm moves against the spring force causing the outlet cross-sectional area at the valve seat to be reduced or closed entirely. The pressure is therefore regulated by the volume flowing through.

The pressure limiting valves are used on the upstream side of an air compressor to ensure the receiver pressure is limited, for safety, and that the supply pressure to the system is set to the correct pressure. Thus pressure limiting valves are used mainly as safety valves and are sometimes referred to as **pressure relief valve**s. They prevent the maximum permissible pressure in a system from being exceeded. If the maximum pressure has been reached at the valve inlet, the valve outlet is opened and the excess air pressure exhausts to atmosphere. The valve remains open until it is closed by the built-in spring after reaching the preset system pressure.

The sequence valve senses the pressure of any external line and compares the pressure of the line against a preset adjustable value, creating a signal when the preset limit is reached.

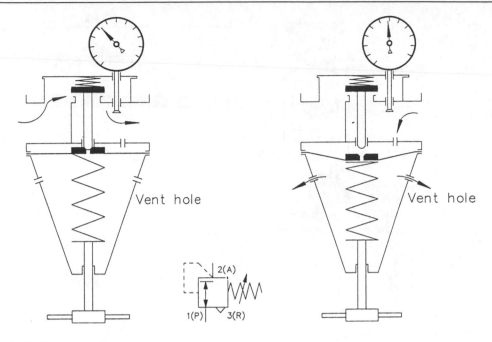

Figure 7.1 Pressure regulating valve

7.2 Pressure control valves – symbols

Referring to Figures 7.2 and 7.3, if the pressure exceeds that set on the spring, the valve opens. The air will then flow from port 1 to port 2. The outlet port 2 will only open if a preset pressure has built up in the pilot line 12. A pilot spool opens the passage from port 1 to port 2.

Figure 7.2 Pressure sequence valve

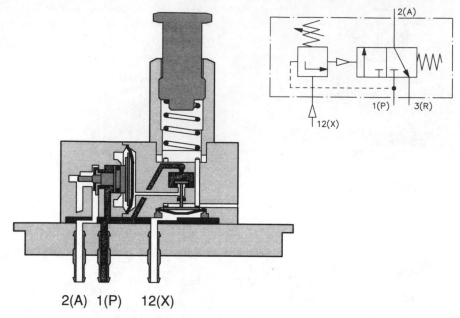

Figure 7.3 Adjustable pressure sequence valve

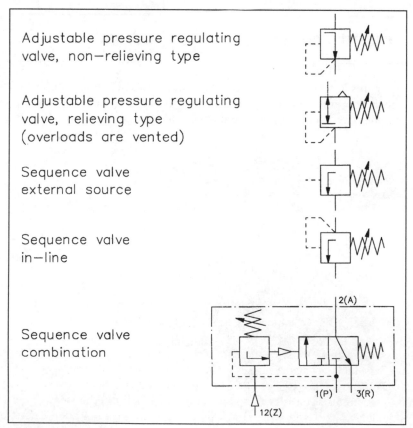

Figure 7.4 Pressure valves

To remind you of the symbols used for pressure control valves, they are shown again in Figure 7.4. Note that the symbols represent the pressure valve as a single position valve with a flow path that is either open or closed initially. In the case of the pressure regulator the flow is always open, whereas the pressure sequence valve is closed until the pressure reaches the limiting value as set on the adjustable spring.

7.3 Combinational valves

The combined functions of various elements can produce a new function. The new component can be constructed by the combination of individual elements or manufactured in a combined configuration to reduce size and complexity. An example is the timer which is the combination of a one-way flow control valve, a reservoir (or accumulator) and a 3/2 way directional control valve.

Other combinational valves include the one-way flow control, two-hand start valve and the impulse valve.

Time delay valve

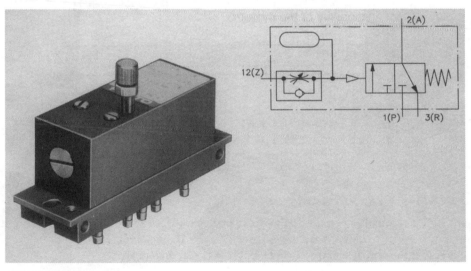

Figure 7.5 Time delay valve

Time delay valve: normally closed

Referring to Figure 7.6, air is supplied to port 1. Control air flows into the valve at port 12. It flows through a one-way flow control valve and, depending on the setting of the throttling screw, a greater or lesser amount of air flows per unit time into the air reservoir. When the necessary control pressure has built up in the air reservoir, the pilot spool of the 3/2-way valve is moved downwards. This blocks the passage of air between ports 2 and 3. The time required for pressure to build up in the air reservoir is equal to the control time delay of the valve.

If the time delay valve is to switch to its initial position, the pilot line 12 must be exhausted. The air flows from the air reservoir to atmosphere through the bypass of the one-way flow control valve and then to the exhaust line. The valve spring returns the pilot spool and the valve disc seat to their initial positions. The working line 2 exhausts to port 3 and port 1 is blocked.

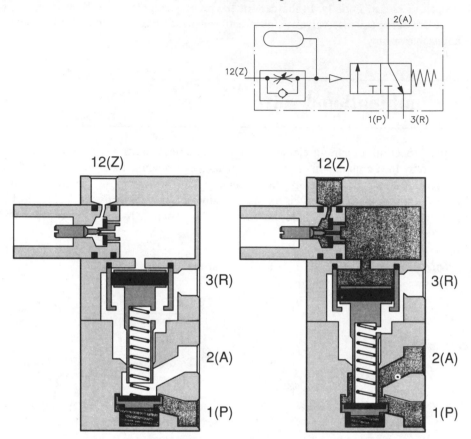

Figure 7.6 Time delay valve: normally closed

Time delay valve: normally open

In Figure 7.7, the normally open time delay valve includes a 3/2-way valve which is open. Initially, the output port 2 is active. When the valve is switched by 10, the output port 2 is exhausted. Thus the result is that the output signal is switched off after a set time delay.

7.4 Solenoid valves

A solenoid consists of a coil of insulated wire, usually in tubular form, which when energised with electric current will produce a magnetic field thereby creating physical movement of an iron rod, known as an armature. Figure 7.8 shows the principle of the solenoid.

The movement of the iron rod may be used as an actuator in order to control pneumatic and hydraulic valves. Where the armature is directly coupled to the valve spool this is known as 'direct acting'. In this case the solenoid would have to generate enough magnetism to move the spool against its seals and possibly against fluid pressure. More usually, the solenoid will operate a pilot air valve which in turn sends an air signal pressure to operate the valve spool. In this way less magnetism and hence less power is required by the solenoid. This is known as 'indirect acting'.

Solenoid coils are available in a variety of a.c. and d.c. voltages, but are often rated in watts or VA because the current flow is dependent upon the voltage and impedance of the coil. Typical solenoid operating voltages are 110 V a.c./d.c., 24 V d.c. and 12 V d.c.

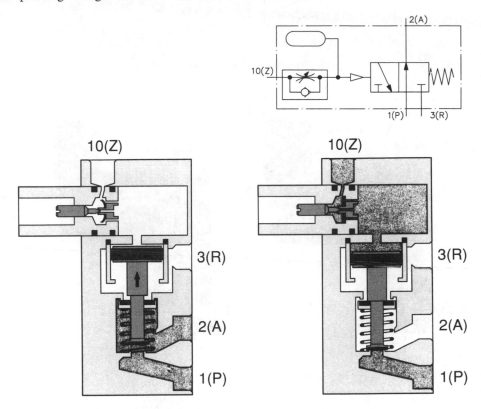

Figure 7.7 Time delay valve: normally open

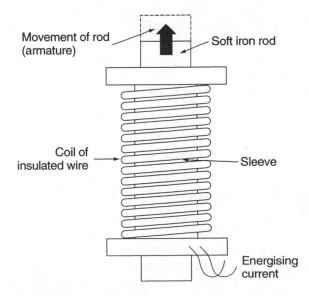

Figure 7.8 Principle of the solenoid

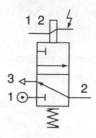

Figure 7.9 Symbol for a 3/2 way valve with direct acting solenoid

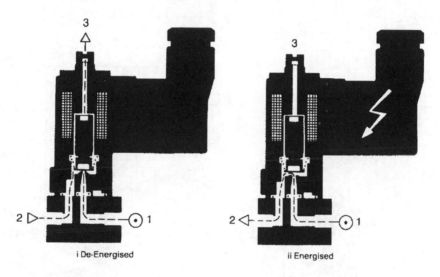

i De-Energised   ii Energised

Figure 7.10 3/2-way pilot operated solenoid valve

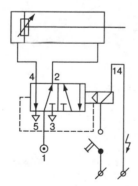

Figure 7.11 5/2 way valve with solenoid and air operation.

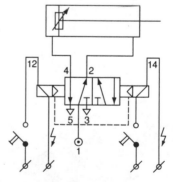

Figure 7.12 5/2 way valve with double-acting solenoid and air operation

NOTE: In all cases the symbol ⚡ is used to illustrate an electrical line

A BS/ISO symbol for a 3/2-way valve operated by a direct-acting solenoid is shown in Figure 7.9. The use of solenoid valves has become more widespread due to the rapid growth of the application of programmable electronic controllers and computer control of fluid power systems (refer to Chapter 10). In Figure 7.10 a 3/2-way pilot solenoid valve is shown in both the energised and de-energised states.

There are several variations on solenoid operation. The 5/2-way valve shown in Figure 7.11 is interesting since it shows mains air biasing the valve to the right, but operation of the solenoid will allow a pilot signal to move it to the left.

The double-acting solenoid pilot operation on a 5/2-way valve is shown in Figure 7.12.

Some advantages of solenoid valves are:

- the speed of transmission of electrical signals is much faster than that of pneumatic and hydraulic signals and can occur over longer distances
- modern electrical components tend to be less expensive and use less space than pneumatic and hydraulic components.

Two disadvantages are:

- the increased complexity of circuits having several distinct power sources (air, hydraulics and electricity) and the need to obtain and maintain a suitable electrical supply
- safety considerations in using electricity in damp, combustible and explosive environments.

8 Actuators

Aims

At the end of this chapter you should be able to:

1 *Appreciate the distinction between linear and rotary motion actuators.*
2 *Understand the principle of operation of various linear motion actuator devices.*
3 *Recognise the various graphical symbols used for linear motion actuators.*
4 *Understand the principle of operation of various rotary motion actuator devices.*
5 *Recognise the various graphical symbols used for rotary motion actuators.*
6 *Be aware of the various methods of visual indication of actuation.*

8.1 Types of actuators

An actuator is an output device for the conversion of supply energy into useful work. The output signal is controlled by the control system, and the actuator responds to the control signals via the final control element. There are, of course other types of output device whose function is to indicate the status of control systems or actuation.

The fluid power actuator can be described under two groups, linear or rotary:

- **linear motion**
  - **single-acting cylinders**
  - **double-acting cylinders**
- **rotary motion**
  - **air motor and hydraulic motor**
  - **rotary actuator.**

8.2 Single-acting cylinder

With single-acting cylinders fluid is applied on only one side of the piston face. The other side is open to atmosphere. The cylinder can produce work in only one direction. The return movement of the piston is effected by a built-in spring or by the application of an external force. The spring force of the built-in spring is designed to return the piston to its start position with a reasonably high speed under no load conditions.

The single-acting cylinder (Figure 8.1) has a single piston seal which is fitted to the fluid power supply side. The exhaust air on the piston rod side of the cylinder is vented to atmosphere through an exhaust port. If this port is not protected by a gauze cover or filter, then it is possible that the entry of dirt particles may damage internal seals. Additionally, a blocked vent will restrict or stop the exhausting air during forward motion, and the motion will be jerky or may stop. Sealing is by a flexible material that is embedded in a metal or plastic piston (Perbunan). During motion, the sealing edges slide over the cylinder-bearing surface.

There are various designs of single-acting cylinders, including:

- diaphragm cylinder
- rolling diaphragm cylinder.

With the diaphragm cylinder construction the friction during motion is less and there is no sliding motion. They are used in short-stroke applications, for clamping, embossing and lifting operations.

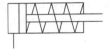

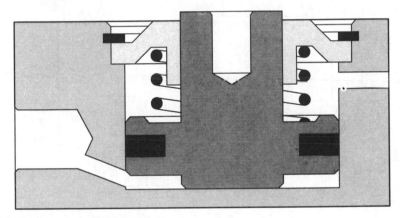

Figure 8.1 Single-acting cylinder

Direct control of a single-acting cylinder

In the example below the piston rod of the single-acting cylinder advances when a button is operated. On release of the button the piston returns to the initial position. A normally closed 3/2-way valve is required for this direct control.

Figure 8.2 Direct control of a single acting cylinder

When the 3/2-way valve is actuated, compressed air flows from 1(P) to 2(A) and the exhaust port 3(R) is blocked. The cylinder extends. When the button is released the valve return spring operates and the cylinder chamber is exhausted through the 3(R) port with the compressed air connection 1(P) blocked. The cylinder retracts under spring force.

Indirect control of a single acting cylinder

With indirect control of a cylinder the 3/2-way control valve is piloted by the 3/2-way push-button valve 1.2. In this way the final control element can be of large orifice size. The control valve then matches the cylinder bore size and flow rate requirements. The push-button valve 1.2 indirectly extends the cylinder 1.0 via the final control element 1.1. In Figure 8.3 when the push button is pressed, the signal 12(Z) pilots the final control element to extend the cylinder against spring force. If the button is released, the signal 12(Z) exhausts and the control element returns to the initial position, retracting the cylinder.

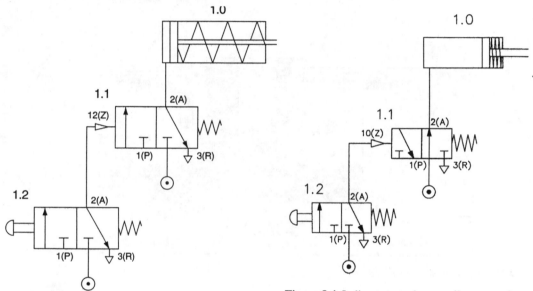

Figure 8.3 Indirect control, normally closed valve

Figure 8.4 Indirect control, normally open valve

In the case of a normally open 3/2-way valve being used as shown in Figure 8.4, the cylinder is initially extended and when the push button is operated the cylinder retracts under the spring force. The final control element 1.1 is not switched at the rest position. The cylinder is pressurised whilst at rest in the initial position. The circuit is drawn with the valve 1.2 unactuated and with the cylinder initially extended due to the signal 2(A) being active at valve 1.1.

8.3 Double-acting cylinder

The construction principle of a double-acting cylinder (Figure 8.5) is similar to that of a single-acting cylinder. However, there is no return spring, and the two ports are used alternatively as supply and exhaust ports. The double-acting cylinder has the advantage that the cylinder is able to carry out work in both directions of motion. Thus installation possibilities are universal. The force transferred by the piston rod is somewhat greater for the forward stroke than for the return stroke as the effective piston surface is reduced on the piston rod side by the cross-sectional area of the piston rod. The cylinder is under control of the supply fluid in both directions of motion.

In principle, the stroke length of the cylinder is unlimited, although buckling and bending of the extended piston rod must be allowed for. As with the single-acting cylinder, sealing is by means of pistons fitted with sealing rings or diaphragms.

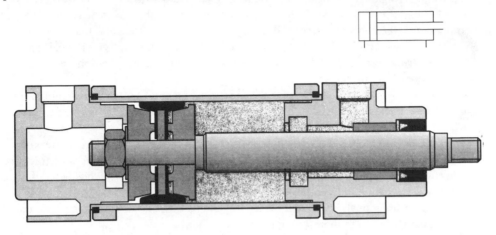

Figure 8.5 Double-acting cylinder

Fluid power cylinders have developed in the following ways:

- contactless sensing requirements – hence the use of magnets on pistons for reed switch operation.
- stopping heavy loads through clamping units and external shock absorbers.
- rodless cylinders where space is limited.
- alternative manufacturing materials such as plastic.
- protective coatings against harsh environments, i.e. acid resistant.
- increased load carrying capacity.
- robotic applications with special features such as non-rotating piston rods, hollow piston rods for vacuum suction cups.

8.4 Cushioned double-acting cylinder

If large masses are moved by a cylinder, cushioning is used in the end positions to prevent sudden damaging impacts (Figure 8.6). Before reaching the end position, a cushioning piston interrupts the direct flow path of the fluid to be exhausted. Instead a very small and often adjustable exhaust aperture is open. For the last part of the stroke the cylinder speed is progressively reduced. If the passage adjustment is too small, the cylinder may not reach the end position due to the blockage of fluid.

When the piston reverses, fluid flows without resistance through the return valve into the cylinder space. With very large forces and high accelerations extra measures must be taken such as external shock absorbers to assist the load deceleration. When cushioning adjustment is being carried out, it is recommended that in order to avoid damage, the regulating screw should first be screwed in slowly to the optimum value.

It is important to consider fitting a magnet to the cylinder piston. Once manufactured, the cylinder cannot normally be fitted with sensor magnets due to the difference in construction.

Construction

The cylinder consists of a cylinder barrel, bearing and end cap, piston with seal (double-cup packing), piston rod, bearing bush, scraper ring, connecting parts and seals.

The cylinder barrel is usually made of seamless drawn steel tube. To increase the life of the sealing components, the bearing surfaces of the cylinder barrel are precision-machined. For special applications, the cylinder barrel can be made of aluminium, brass or steel tube with hard-chromed bearing surface. These special designs are used where operation is infrequent or where there are corrosive influences.

The end cap and the bearing cap are, for the most part, made of cast material (aluminium or malleable cast iron). The two caps can be fastened to the cylinder barrel by tie rods, threads or flanges.

The piston rod is preferably made from heat-treated steel. A certain percentage of chrome in the steel protects against rusting. Generally the threads are rolled to reduce the danger of fracture.

A sealing ring is fitted in the bearing cap to seal the piston rod. The bearing bush guides the piston rod and may be made of sintered bronze or plastic-coated metal.

In front of this bearing bush is a scraper ring. It prevents dust and dirt particles from entering the cylinder space. Bellows are therefore not normally required.

The materials for the double-cup packing seals are:

- Perbunan, for –20°C to +80°C
- Viton,   for –20°C to +190°C
- Teflon,   for –80°C to +200°C

O-rings are normally used for static sealing.

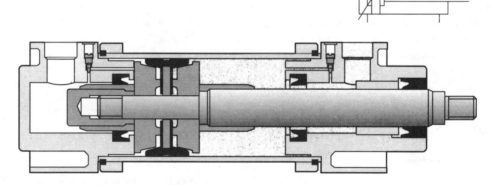

Figure 8.6 Cushioned double-acting cylinder

8.5 Linear actuators – symbols

To remind you of the symbols they are reproduced in Figure 8.7. Note that the single-acting cylinder and the double-acting cylinder form the basis for design variations. The use of cushioning to reduce loads on the end caps and mountings during deceleration of the piston is important for long-life and smooth operation. The cushioning can be either fixed or adjustable. The cushioning piston is shown on the exhaust side of the piston. The arrow indicates adjustable cushioning and not the direction of cushioned motion.

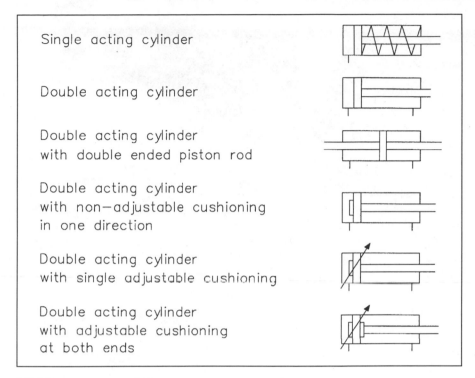

Figure 8.7 Linear actuator symbols

8.6 The rodless cylinder

This double-acting linear actuator (cylinder without piston rod) consists of a cylindrical barrel and rodless piston (Figure 8.8). The piston in the cylinder is freely movable according to actuation, but there is no positive external connection. The piston is fitted with a set of annular permanent magnets. Thus, a magnetic coupling is produced between the slide and the piston. As soon as the piston is moved by fluid the slide moves synchronously with it. The machine component to be moved is mounted on the carriage. This design of cylinder is specifically used for extreme stroke lengths of up to 10 m. An additional feature of the rodless design is the flat bed mounting available on the carriage as opposed to the threaded piston rod type of construction.

Control of a rodless cylinder

For the accurate positioning of the carriage, the circuit for the rodless cylinder uses check valves to prevent the carriage from creeping. Referring to the circuit, the push button for the carriage to move to the right is the right-hand valve 1.2. In this case the valve that exhausts the air controls the motion of the cylinder.

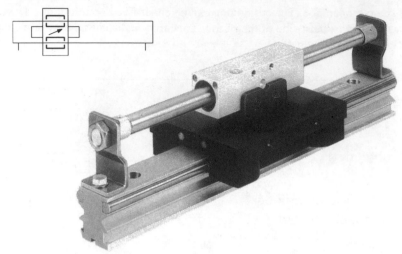

Figure 8.8 Rodless cylinder

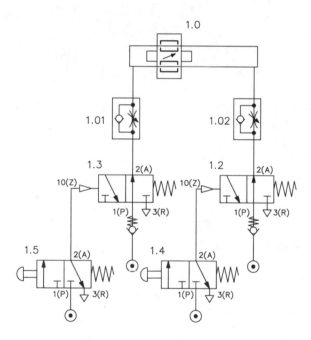

Figure 8.9 Control of a rodless cylinder

8.7 Hydro-pneumatic systems

Since air is compressible and springy, it is extremely difficult to produce a very slow, steady movement by means of an air cylinder. Similarly, 'inching' cannot be achieved easily. However, these problems may be solved by adding a simple hydraulic system. Figure 8.10 shows a combination of air cylinder and hydraulic dash-pot consisting of two cylinders of equal stroke having a common piston rod. The front cylinder is filled with oil and the two sides of its piston have equal areas. When air pressure (P_1 or P_2) is applied to the rear cylinder, the pistons move together displacing oil through the bypass from one side of the front piston to the other. In Figure 8.10(A) a variable restrictor is fitted into the bypass; the movement obtained is steady and its speed can be varied by adjusting the restrictor. In Figure 8.10(B) the restrictor has been replaced by a simple on/off valve, and since oil is incompressible this produces a neat 'inching' system.

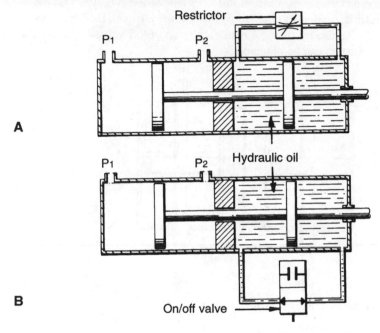

Figure 8.10 Air cylinder and hydraulic dash-pot arrrangement. The valve in B is an on/off valve

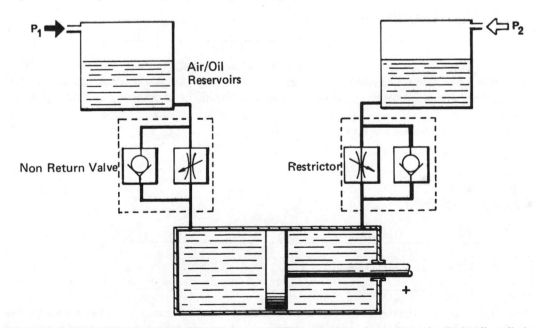

Figure 8.11 Obtaining slow, steady motion using two air/oil reservoirs and a double acting hydraulic cylinder

An alternative method of obtaining slow, steady motion is shown in Figure 8.11. Compressed air is supplied to one or other of two air/oil reservoirs which are connected to a double-acting hydraulic cylinder through variable restrictors bypassed on the exhaust stroke by non-return valves. This system allows different speeds in the two directions.

Figure 8.12 shows how, without being linked mechanically, two air-powered cylinders can be piped up to move together through equal distances. Pressure P_1 applied to cylinder **A** produces movement of the piston and displaces oil from the front half of the cylinder. This displaced oil is fed to the rear of cylinder **B**, producing equal movement of the piston in **B**. The motion is reversed by applying pressure P_2. Again, by fitting a restrictor or an on/off valve in the oil line, various types of controlled movements can be obtained.

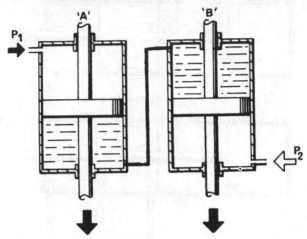

Figure 8.12 Two air-powered cylinders hydraulically linked

8.8 Air motors – general

Devices which transform pneumatic energy into mechanical rotary movement with the possibility of continuous motion are known as air (pneumatic) motors. The air motor with unlimited angle of rotation has become one of the most widely used working elements operating on compressed air systems.

Air motors are categorised according to design:

- piston motors
- sliding vane motors
- gear motors
- turbines (high flow).

8.9 Piston motors

This type of design is further subdivided into radial and axial piston motors (Figure 8.13). The crank shaft of the motor is driven by the compressed air via reciprocating pistons and connecting rods. To ensure smooth running several pistons are required. The power of the motors depends on input pressure, number of pistons, piston area, stroke and piston speed.

The working principle of the axial piston motor is similar to that of the radial piston motor. The force from five axially arranged cylinders is converted into a rotary motion via a swash plate. Compressed air is applied to two pistons simultaneously, the balanced torque providing smooth running of the motor. These pneumatic motors are available in clockwise or anticlockwise rotation. The maximum speed is around 5000 rpm, the power range at normal pressure being 1.5–19 kW (2–25 hp).

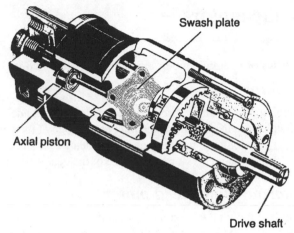

Figure 8.13 Axial piston motor

8.10 Sliding vane motors

Because of their simple construction and low weight, sliding vane motors are used for hand tools. The principle of operation is similar to the sliding vane compressor.

An eccentric rotor is contained in bearings in a cylindrical chamber. Slots are arranged in the rotor. The vanes are guided in the slots of the rotor and forced outwards against the inner wall of the cylinder by centrifugal force. This ensures that the individual chambers are sealed.

The rotor speed is between 3000 and 8500 rpm. Here too, clockwise or anticlockwise units are available as well as reversible units. Power range 0.1–17 kW (0.1–24 hp).

8.11 Pneumatic gear motors

In this design, torque is generated by the pressure of the air against the teeth profiles of two meshed gear wheels. One of the gear wheels is secured to the motor shaft. These gear motors are used in applications with a very high power rating (44 kW/60 hp). The direction of rotation is also reversible when spur or helical gearing is used.

8.12 Turbines (flow motors)

Turbine motors can be used only where a low power is required. The speed range is very high. For example, dentists' air drills operate at 500 000 rpm. The working principle is the reverse of the flow compressor.

8.13 Hydraulic motors

A number of different types of hydraulic motor exist, the following two will be discussed:

Hydraulic gear motor

This consists of two meshing gears which rotate in opposite directions inside a housing with an inlet and an outlet port. Hydraulic system pressure at the inlet acts on one of the gears creating an imbalance which results in gear rotation. The meshing gear is attached to a shaft which supplies torque to a load. Increasing hydraulic system pressure or the size of the gear teeth increases torque output.

Hydraulic high torque, low speed motor

In this type of design of hydraulic motor a rotary commutator valve is used to feed hydraulic oil to fluid chambers in the rotor set in such a way that an inbuilt speed and torque variation is achieved.

Both the gear motor and high torque, low speed motors are usually very compact in design as shown in Figures 8.14 and 8.15.

Figure 8.14 Hydraulic gear motor

Figure 8.15 Hydraulic high torque, low speed motor

8.14 Summary of characteristics of air and hydraulic motors

- smooth regulation of speed and torque
- small size (weight)
- overload safe
- insensitive to dust, water, heat, cold
- explosion proof
- large speed selection
- minimal maintenance
- direction of rotation easily reversed.

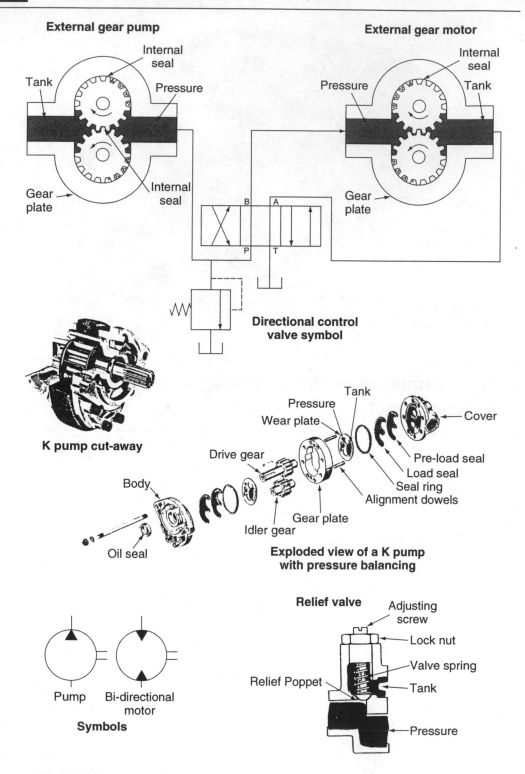

External gear pump

Internal seal

Tank

Pressure

Gear plate

Internal seal

External gear motor

Internal seal

Pressure

Tank

Gear plate

B   A

P   T

Directional control valve symbol

K pump cut-away

Tank

Pressure

Wear plate

Drive gear

Body

Oil seal

Idler gear

Gear plate

Cover

Pre-load seal

Load seal

Seal ring

Alignment dowels

Exploded view of a K pump with pressure balancing

Pump   Bi-directional motor

Symbols

Relief valve

Adjusting screw

Lock nut

Valve spring

Relief Poppet

Tank

Pressure

Figure 8.16 Basic assembly and operation of gear pumps and motors

8.15 Rotary actuators

Salient design features of rotary actuators (Figure 8.17) are:

- small and robust
- precision machined and hence very efficient
- available with contactless sensing
- adjustable for angular displacement
- constructed from lightweight material
- easy to install.

The compact rotary actuator is suited to robotics and materials handling applications where there is limited space.

8.16 Rotary motion actuators – symbols

Figure 8.17 Rotary actuator

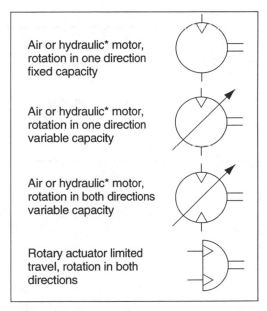

Air or hydraulic* motor, rotation in one direction fixed capacity

Air or hydraulic* motor, rotation in one direction variable capacity

Air or hydraulic* motor, rotation in both directions variable capacity

Rotary actuator limited travel, rotation in both directions

Figure 8.18 A summary of the basic rotary motion actuator symbols

*__Note:__ Hydraulic devices will have ▽ shown filled in thus ▼, so a hydraulic motor having rotation in one direction and fixed capacity would be indicated as

8.17 Methods of visual indication of actuation

Some of the visual devices are:

- optical indicators, single and multiple coloured units
- pin type optical indicators, for visual display and tactile sensing
- counters, for displaying counting cycles
- pressure gauges, to indicate air and hydraulic oil pressure values
- timers, with visual indication of time delay.

With the optical indicators the colour codes represent certain functions in the cycle. The visual indicators are mounted on a control panel to indicate operational status of the control functions and the sequential steps currently active. Typical colours for visual indicators are:

Colour	Meaning	Notes
Red	Immediate danger, alarm	Machine status or situations requiring immediate intervention (No entry)
Yellow (Amber)	Caution	Change or imminent change of conditions
Green	Safety	Normal operation, safe situation, free entry
Blue	Special information	Special meaning which cannot be made clear by red, yellow, amber or green
White (Clear)	General information	Without special meaning. Can also be used in cases where there is doubt as to the suitability of the colours, red, yellow, amber or green

9 Pneumatic and Hydraulic Circuits and Arrangement of Components

Aims

At the end of this chapter you should be able to:

1 *Understand the basic AND/OR logic functions when applied to fluid power circuits.*
2 *Appreciate the concept of a latched (memory) circuit.*
3 *Understand the operation of simple sequential circuits.*
4 *Appreciate the use of the Cascade technique when used to switch groups of valves in order to avoid trapped air signals.*

9.1 The logic OR function

A logic OR function requires at least **one** input device to be initiated in order to cause actuation of the output.

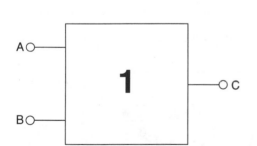

Truth table

Input		Output
A	B	C
0	0	0
1	0	1
0	1	1
1	1	1

9.2 The fluid power logic 'OR' circuit

The pneumatic logic OR function can be achieved using a **shuttle valve** as shown in the circuit diagram in Figure 9.1.

The shuttle valve is connected to the junction between the two 3/2-way push-button valves. Upon operation of one of the push-buttons, a signal is generated at the X or Y side of the shuttle valve. This signal passes through the shuttle valve and is emitted at port A. This reverses the control valve via pilot port 14(Z), and the cylinder extends. The control valve can be a 4/2-way or a 5/2-way valve and can be sized to suit the flow rate required for the cylinder speed. If both of the signals produced via the push-button valves are removed, then the shuttle valve will release the 14(Z) pilot signal back through the exhaust port of one of the 3/2-way valves. The return spring in the control valve switches the 5/2-way valve to the initial position. The outlet 2(B) is active with the outlet 4(A) exhausted to atmosphere, and the cylinder retracts.

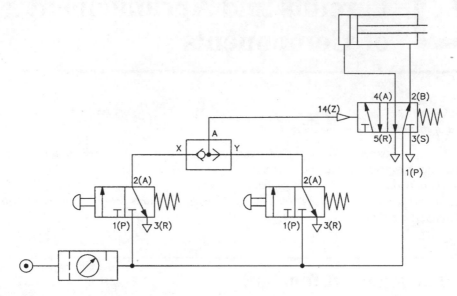

Figure 9.1 Fluid power logic OR circuit

9.3 The logic AND function

A logic AND function requires **all** input devices to be initiated simultaneously in order to cause actuation of the output.

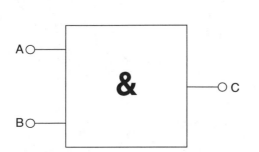

Input		Output	Truth table
A	**B**	**C**	
0	0	0	
1	0	0	
0	1	0	
1	1	1	

9.4 The fluid power logic AND circuit

The pneumatic logic AND function can be achieved using a **two-pressure valve** as shown in the circuit diagram in Figure 9.2.

The two-pressure valve is connected to the junction between the two 3/2-way push-button valves. Upon operation of one of the push-buttons, a signal is generated at the X or Y side of the two-pressure valve. This signal is blocked by the two-pressure valve. If the second pushbuttton is also operated, then the two-pressure valve will produce a signal at port A which operates the control valve 14(Z) pilot signal against the spring return and the cylinder extends.

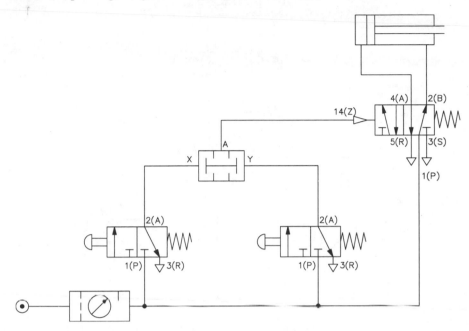

Figure 9.2 Fluid power logic AND circuit

9.5 The latched (memory) circuit

In this arrangement an output device will remain actuated until the original fluid power signal is 'cancelled' and a further signal is provided to actuate the output device in a different mode.

Take for example the circuit arrangement in Figure 9.3. Upon operation of push-button A, a signal is generated at the 2(A) port and the pilot port 14(Z) of valve C. The 5/2-way memory valve switches and the signal from port 4(A) fully extends the cylinder F. If the push-button valve A is released, the signal at 14(Z) is exhausted at the 3(R) port of the push-button valve A. The valve C remains in the switched position until the push-button valve B is operated. If the push-button valve A is released and therefore there is no signal at 14(Z) then the signal generated by B will return the memory valve to its initial position and the cylinder retracts. The cylinder remains retracted until a new signal is generated at 14(Z) by the valve A. The cylinder piston rod will extend and retract if there are no obstructions, but there is no confirmation that the cylinder is in its fully extended position. If both the 14(Z) signal and the 12(Y) signal are active due to both push-buttons being operated, then the memory valve will remain in the last position.

The flow control valves D and E have been fitted to throttle the exhausting air in both directions of piston motion. The supply air is transferred through the bypass check valve of the flow control valves, giving unrestricted supply to the cylinder.

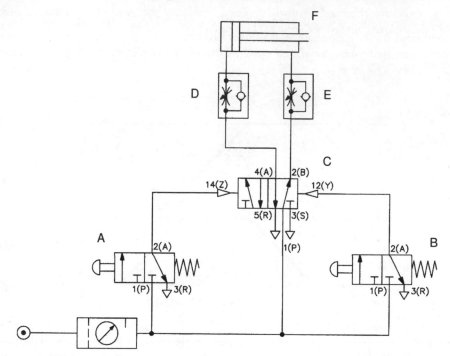

Figure 9.3 Fluid power latched (memory) circuit using a 5/2-way directional control valve

9.6 Sequential circuits

Sequential circuits are often used for the control of multiple actuators. Take for example the circuit in Figure 9.4.

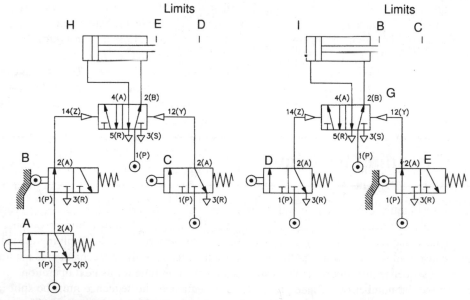

Figure 9.4 Circuit for two actuator sequential operation

In this circuit the sequential task is dependent on roller-operated limit valves B, C, D , E.

The start conditions for the control are that cylinder H is retracted and the start-button A is operated. The process can be broken down into steps as shown on the displacement-step diagram, Figure 9.5.

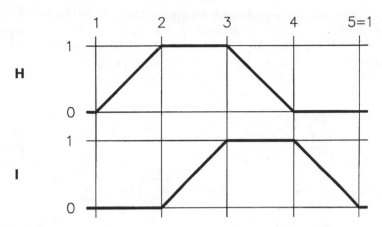

Figure 9.5 Displacement-step diagram

When looking at the placement of the roller valves, the current action and then reaction must be analysed. These will determine the valve positioning. The sequence of the actuators can be described as: Cylinder H out, I out, H back, I back.

The notation for this is sometimes described as H+, I+, H– and I–. The sequence of operation is:

- Operate push-button A
- cylinder H outstroking   →   limit switch on E operated
- cylinder H fully outstroke →   limit switch on D operated
- cylinder I outstroking   →   limit switch on B operated
- cylinder I fully outstroke →   limit switch on C operated
- cylinder H instroking   →   limit switch on D operated
- cylinder H fully instroke   →   limit switch on E operated
- cylinder I instroking   →   limit switch on C operated
- cylinder I fully instroke   →   limit switch on B operated
- cycle complete
- commence again by pushing button A.

9.7 Cascading techniques

The cascade technique is to switch on and off the supply air to the critical trip valves in groups. The need for this will occur when a trip valve's mechanism is still held down, but the output signal has been used and requires removing. By switching off the group air that is supplying the valve, the output is also removed and achieves the desired result. After the valve's mechanism is naturally released in the sequence, the group supply is switched on again in time for its next operation.

To determine the number of cascade groups for any sequence, the sequence must be split into groups starting at the beginning, so that no letter occurs more than once in any group. The group numbers are given Roman numerals to avoid confusion with other numbering systems that may exist on valve arrangements, particularly on larger systems. The placing of the run/end valve should be in the line that selects group I. This determines that the first task of group I is to signal the first movement of the sequence. In addition, when the circuit is at rest, inadvertent operation of an uncovered trip valve will not risk an unwanted operation of a cylinder.

By studying Figure 9.6. it can be seen that the sequence splits into two groups. These groups are supplied from a single double pressure-operated 5/2-way valve, so that only one group can exist at any one time. This is known as the cascade valve. It can also be seen that neither of the 5/2-way valves controlling the cylinders can have the outstroke (+) and instroke (−) command lines as opposed signals, since their source is from different groups.

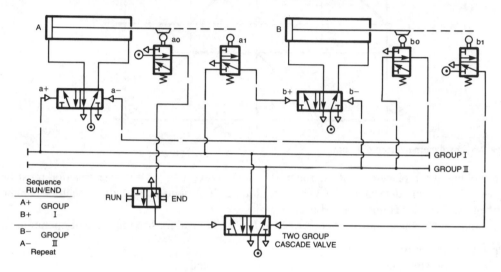

Figure 9.6 Two group cascade system

The circuit can be traced as follows:

```
To start, set run/end valve to 'run', generating a command to select group I
Group I gives a command a+. Cylinder A moves a+.
Valve a1 is operated and generates a command b+.
Cylinder B moves +.
Valve b1 is operated and generates a command to select group II.
Group II gives a command b− (because group I has been switched off there is no
opposing signal from a1).
Cylinder B moves −.
Valve b0 is operated and generates a command a− (no opposed signal).
Cylinder A moves −.
Valve a0 is operated and generates a command to start the sequence again.
If at any time the run/end valve is switched to 'end', the current cycle will
be completed, but the final signal will be blocked and no further operation will
occur.
```

The rules for interconnection are as follows:

1 The first function in each group is signalled directly by that group supply.
2 The last trip valve to become operated in each group will be supplied with main air and cause the next group to be selected.

3 The remaining trip valves that become operated in each group are supplied with air from their respective groups and will initiate the next function.

Sequence splitting into three or more groups requires cascade valves to be interconnected. The patterns shown in Figures 9.7 and 9.8 can be used.

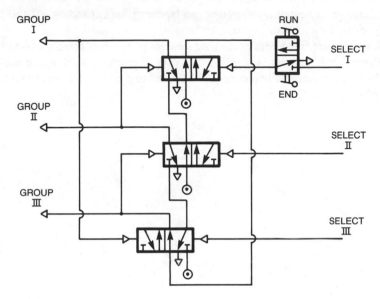

Figure 9.7 Three-group cascade system

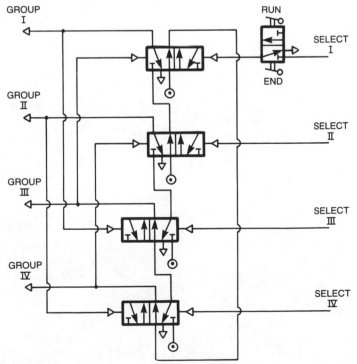

Figure 9.8 Four-group cascade system

With the exception of a two-group system, the cascade sub-circuits or building blocks use one 5/2-way valve for each of the total number of groups. Cascade sub-circuits can be constructed using one less 5/2-way valve than the number of groups, or even an equal number of 3/2-way valves, but both are less secure against inadvertently generated signals. It can be seen from Figures 9.7 and 9.8 that the four-group system is simply an expansion of the patterns for the three-group system. A pattern for any number of groups can be drawn by expanding out the pattern with an extra 5/2-way valve for each extra group.

In general, the current trend with pneumatic systems of medium complexity and upwards, is to use an electro-pneumatic solution. For this, 5/2-way double solenoid valves are used to control the cylinders and a micro-electronic sequencer or PLC used to control the logic. Reference to this type of control is illustrated in Chapter 10.

10 Electro-pneumatics and Electro-hydraulics

Aims

At the end of this chapter you should be able to:

1 *Appreciate the use of electronics in fluid power systems.*
2 *Appreciate how programmable logic controllers can be used to control fluid power systems.*

Programmable electronic control equipment

10.1 Electro-pneumatics

In recent years medium to complex industrial pneumatic systems have been totally transformed in the manner in which they work. The previously all-pneumatic methods have now been replaced by the combined use of electronics for the control and sequencing of applications. Only systems of low complexity and those in use in hazardous areas, not compatible with electronics, remain as pure pneumatic systems.

A purely pneumatic system can be viewed as three main sections:

1 Generation and preparation of the compressed air power source.
2 Power actuation of pneumatic cylinders through directional control valves.
3 Pneumatic signal processing or logic control.

The influence of electronics can be seen in all of these sections. For example, in section 1 there is electronic management control of compressors and electronically controlled pressure regulation. In section 2 there are solenoid valves that provide proportional flow and proportional pressure, together with air cylinders having electronic proportional feedback. In section 3, however, for many systems pneumatic logic has been completely replaced by electronic sequence or logic control.

Programmable sequence controllers (sequencers) and programmable logic controllers (PLCs) are used to control the sequence of actions in a special-purpose machine cycle. These are the most commonly used devices and offer a wide range of features such as timing, counting, looping and logic functions.

To decide when to use electronics to control a fixed sequence of events, simply take the sequence and apply the cascade technique of splitting it into groups. If the sequence is not very complicated, it will result in a two or three-group pneumatic cascade system. Also, if there are only a few time delays and no long counting procedures, then the purely pneumatic solution will be lower in cost and likely to be more robust and simple to maintain.

If, however, the sequence is complicated, it is likely to result in four or more groups of cascade. There may also be one or more of the cylinders that operate several times within one overall cycle. Other complications could also be long counting operations, or a number of time delays, requiring a high degree or repeatable accuracy. For these applications the electronic controller will usually be the better choice.

Sequencers and PLCs communicate with the external hardware in a continuous process of command and feedback instructions. See Figure 10.1.

Inputs to the controller indicate the completion of a cylinder movement. These are most conveniently achieved by using a magnetic cylinder fitted with reed switches. The reed switch consists

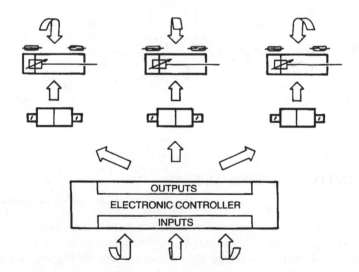

Figure 10.1 The process of command and feedback in programmable electronic control and communication

of two spring-like metal reeds within a sealed enclosure. When the magnet around the piston is within range, the reeds are magnetised, each having a N and S pole. As the free ends will be of opposite polarity they snap together, making contact. When the piston moves out of range, the reeds lose their magnetism and spring apart again. See Figure 10.2.

For environments where there are likely to be strong magnetic fields present, for example close to arc welding transformers, reed switches may become operated inadvertently, therefore mechanical limit switches should be used instead.

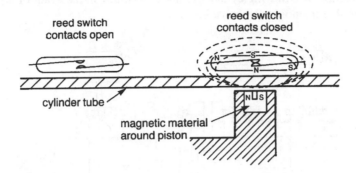

Figure 10.2 Reed switch operation

Outputs from the controller, command cylinder movements and will be connected to solenoid directional control valves. Even though many controllers have some form of built-in suppression it is good practice to provide additional suppression to the solenoid coils. When a coil is switched off the collapsing magnetic field within the windings tries to continue the current flow, and without suppression the back emf would produce a very high negative potential that could cause arcing across the just opened switch. This could easily damage the switch and cause interference.

There are a variety of suppression methods available but the simplest method for dc coils is to connect a diode across the coil terminals. The diode (which is the electronic equivalent of a pneumatic non-return valve) will block the bypass flow of current when the coil is being driven, but when the current is switched off it effectively connects the two ends of the coil together for that particular direction of current flow. The induced current can then flow freely around the coil at the low potential difference of less than 1 volt until the magnetic field has collapsed and the solenoid valve has closed. See Figure 10.3.

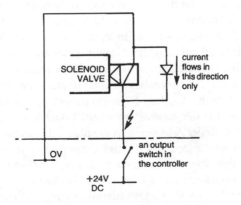

Figure 10.3 Use of diode protection to suppress back EMF on a solenoid valve

One advantage of using electronic control is simplicity and standardisation of installation. See Figure 10.4. This shows a typical layout for a wide range of sequence and logic controllers. Magnetic air cylinders are used and identified as A, B, C and D. The reed switches attached to them are identified as a0, proving the instroked position of A, and a1, proving the outstroked position of A, and likewise for the remaining cylinders B, C and D. Double solenoid valves are chosen so that in the event of an electrical power failure the valves will retain the prevailing state. Any movement that had already been started will be completed, after which the total cylinder state will remain in that position. If a moving cylinder is not to continue in the event of a power failure, a 5/3, all sealed, double solenoid valve must be used.

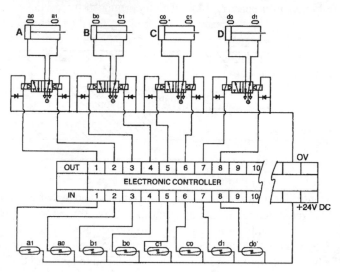

Figure 10.4 Typical layout for sequence and logic control

The controller will have two rows of terminals. One set for inputs, the other for outputs. The outputs, usually at 24 V dc, will each drive through a solenoid coil to return on a common line to the 0 V terminal. This assumes the controller has its own built-in power supply. The inputs receive a signal from their appropriate reed switches which are connected with a common 24 V dc supply. This supply will have limited current, usually just enough to switch on all of the inputs only, so it must not be used for any other purpose.

This standard wiring layout can be the same for any application using four actuators. This will be regardless of the length or complexity of the actual sequence since this is determined by the program given to the controller. The number of independent cylinders that can be used with a particular controller is determined by the number of inputs and outputs available (I/O). In practice the cylinder and valve sizes will be selected according to the duty required of them. Some of the input devices may be from mechanical limit switches or other devices, but these will simply replace the relevant reed switches shown in the diagram. Note the pattern with which the solenoid valves and reed switches have been wired to the number terminals. When output 1 is energised it will operate that solenoid that causes cylinder A to move to the plus or outstroked position. This will result in reed switch a1 turning on which feeds back a signal to input 1. Although the I/O can usually be wired in any chosen order, this, like-number pattern, saves the programmer and the maintenance engineer a great deal of cross-reference work since each numbered output will result in the same numbered input. This will be true for single movements or combinations of movements, occurring at the same time.

Programming methods vary with the type of controller, and someone with no experience will generally find it easier than they think. Sequencers are designed to be easy to program and are a good choice for machines where the actions are performed in a one-after-the-other interlock. Sequencers are able to jump from one part of a sequence to another, run sections of a sequence in a repeating loop, time, count and perform logic functions such as AND, OR, NOT, etc. It may also be possible to hold several sequences in a memory and select the desired one for a particular task. In some models it is even possible to run parallel sequences. Sequencers will have a range of control buttons as a built-in feature. These will provide facilities such as, run/end cycle, emergency stop, single cycle, auto-cycle and manual override.

It takes a little longer to program a PLC. This is produced by keying-in a list of logic statements. These statements are first determined by drawing a ladder diagram. See Figure 10.5. A ladder diagram for a PLC is a logic circuit of the program as it relates to a machine's function and sequence. These are similar to and derived from the ladder electrical circuits used to design electro-mechanical relay systems. For those with no experience, the straightforward and logical nature of this programming method will quickly become apparent. A PLC is ideally suited to machine functions where there are several parallel parts to the sequence.

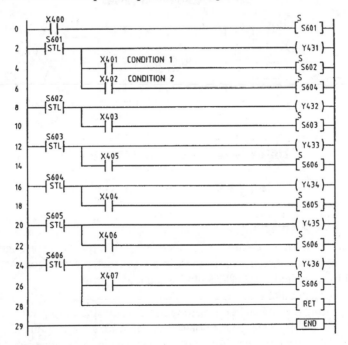

Figure 10.5 A PLC relay logic ladder diagram

Also, where there are a large number of decision-making points or there are process loops to be performed. A PLC will not have any of the push-button controls associated with a sequencer. These have to be added as external hardware and wired to the input terminals. They will then only perform the tasks required, if the logic program has been written to accept them in the way intended.

This chapter has concentrated on electronic control of the typical machine functions, where all movements occur between two fixed limits. Proportional valves and cylinders should also be mentioned as this is a fast-growing field. With the increasing sophistication of machine capability it is sometimes necessary to accurately place the piston of an actuator at various points along its stroke. In addition to this, it is often required to hold station even under changing load conditions. For this, a programmable position controller provides the additional electronics. This will place and hold the actuator's piston rod in a position proportional to a reference signal, or position value. When used within a sequence the reference value can be changed at the appropriate time by the sequencer or PLC. Typical application are the control of the main axes of variable position pick-and-place arms and robotics.

10.2 Electro-hydraulics

Much of what has been said concerning the transformation of purely pneumatic arrangements into hybrid electro-pneumatic systems is equally true in the field of electro-hydraulics. Suffice it to say that significant advances have been made in respect to the speed and reliability of operation of systems by the introduction of electronics. An application of a programmable logic controller to the control of a hydraulic system is shown in Figure 10.6.

Using two solenoid valves

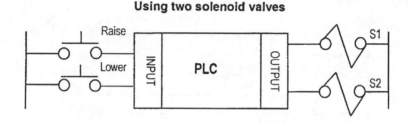

Relay ladder logic arrangements

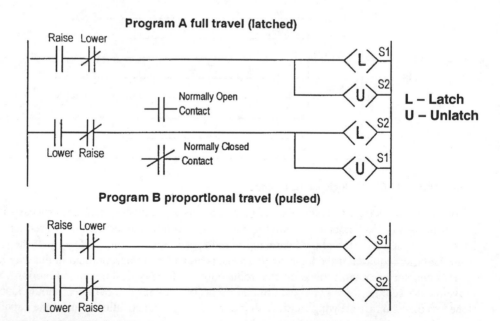

Figure 10.6 Using two solenoid valves, this illustrates the basic method used to interface a PLC to the control of a hydraulically driven warehouse door

11 Fluid Power Measurement Systems

Aims

At the end of this chapter you should be able to:

1 *Recognise the basic elements that constitute a fluid power measuring system.*
2 *Appreciate the function of each of these elements.*
3 *Understand the principle of operation of some common types of pressure, flow and temperature measuring systems.*
4 *Appreciate some common terms applied to measuring systems.*

11.1 The basic fluid power measuring system

In general, fluid power measuring systems can be represented as having three elements:

1 a detecting element known as a **transducer** which produces a signal related to the parameter of the fluid being detected
2 an element called a **signal conditioner** which converts the signal from the transducer into a form which can be displayed and/or recorded
3 a **display** and/or **recording** element which enables the signal to be read and/or recorded.

A block diagram representation of a measuring system is shown in Figure 11.1.

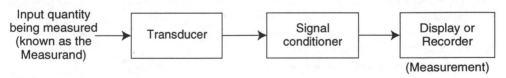

Figure 11.1 Measurement system block diagram

11.2 Pressure measurement

There are several types of device used for pressure measurements in fluid power applications. The following three will be considered:

■ Bourdon tube pressure gauge
■ strain gauge pressure measuring system
■ piezoelectric pressure measuring transducer.

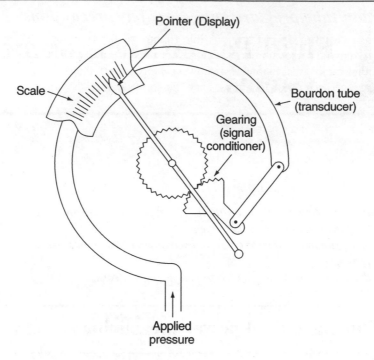

Figure 11.2 Bourdon tube pressure gauge

The Bourdon tube pressure gauge

The Bourdon tube is in effect a **transducer,** whereby an increase in pressure causes the tube to straighten a little. The input of pressure is therefore changed into mechanical displacement of the tube. In this case the transducer changes information about pressure into information in the form of mechanical displacement.

The displacement of the tube is comparatively small and needs to be made larger for display and reading. This is done by gearing, which performs the function of **signal conditioner**. Without the gearing, the output from the Bourdon tube would be too small to be read. In effect the gearing acts as an amplifier.

   The movement of the gear causes a pointer to move across a scale. The pointer and scale constitute what is called the **display** element.

   A block diagram representing a Bourdon tube is shown in Figure 11.3.

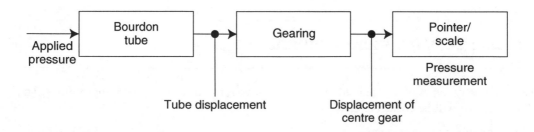

Figure 11.3 Bourdon gauge pressure measurement system block diagram

The Bourdon tube pressure gauge – safety precautions

Precautions must be taken against any possible rupture of the Bourdon tube on high pressure gauges in order to avoid personal injuries. Gauges for pressure above **70 bar** are constructed so as to direct any burst backwards. Figure 11.4 shows a safety pattern gauge.

Note: Any gauge used on oxygen must have the above features and be completely free of oil and grease

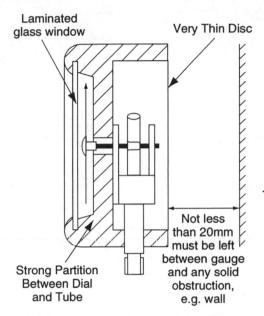

Figure 11.4 Bourdon tube pressure gauge safety pattern glass

The strain gauge pressure measuring system

A strain gauge is in effect a resistive wire of very small diameter (typically 0.01 mm) which when subjected to a strain as a result of applied pressure will cause the electrical resistance of the wire to change in direct proportion to the strain. This property can thus be measured electrically and thereby give a measurement of pressure.

Generally, it is the 'unbonded' strain gauge which is used for pressure measurement. Figure 11.5 shows a typical arrangement. In order to complete the measuring system, the strain gauge may be connected into the arm of a Wheatstone bridge together with a display or recording device, as shown in Figure 11.6. The resulting measurement block diagram is shown in Figure 11.7.

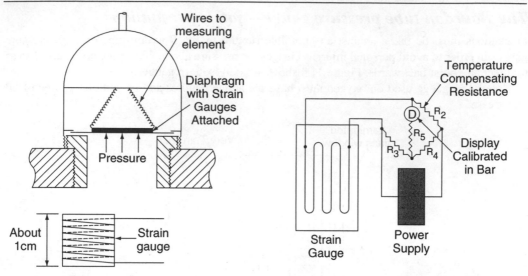

Figure 11.5 Strain gauge pressure sensor

Figure 11.6 Strain gauge/Wheatstone bridge pressure measuring system

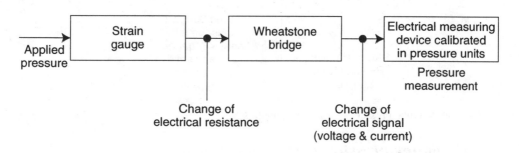

Figure 11.7 Strain guage/wheatstone bridge measurement system block diagram

The piezoelectric pressure transducer

The piezoelectric effect occurs when a force is applied to the opposite parallel faces of certain crystals (for example, quartz, Rochelle salt, barium titanate). This manifests itself as an electric charge between the faces of the crystal and can be used as a measurement of pressure. Recall from Chapter 2 that pressure is force applied to a surface area.

Figure 11.8 illustrates the concept, with Figure 11.9 showing a typical piezoelectric sensor.

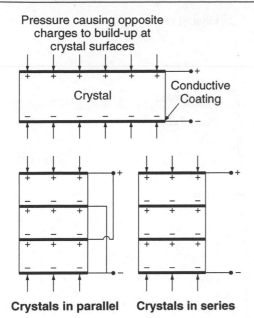

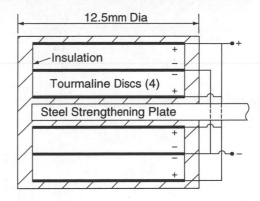

Figure 11.9 Piezoelectric sensor

Crystals in parallel   Crystals in series

Figure 11.8 Piezoelectric crystals

11.3 Flow measurement

Fluid flow occurs in both pneumatic and hydraulic systems. Flow measurement may be concerned with:

- volumetric flow
- mass flow
- or velocity of flow.

It is the velocity of flow which is used in the design of fluid power systems.

The rotameter

The rotameter is probably the simplest method of measuring flow. This is a variable area meter in which a long taper tube is graduated on its vertical axis. A float moves freely in the tube and by an arrangement of shaped flutes in the float it slowly rotates. As flow rate increases, the float rises in the tube so that the annular area increases, which means that the differential pressure across the tube is at a constant value. The float can be arranged with a magnet attachment, and a follower magnet outside will transmit motion to a pointer via linkage if required. Figure 11.10 shows the concept.

The turbine flowmeter

The turbine flowmeter comprises a rotor immersed within the fluid and supported centrally in the pipe. A small permanent magnet can be mounted on the tip of one or more of the rotor blades. A detector coil is used as an electromagnet pick-up to sense the rate of revolution of the rotor. The faster the fluid flows in the pipe, the faster will the rotor rotate and hence this can be used as a means of measuring flow. Figure 11.11 illustrates the concept.

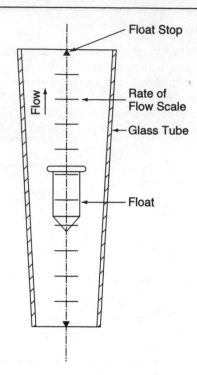

Figure 11.10 Rotameter

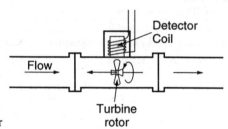

Figure 11.11 Turbine flowmeter

11.4 Temperature measurement

As we have seen in Chapter 5, the temperature of compressed air can vary considerably at different points in its generation, supply and distribution cycle. Similarly, the temperature of hydraulic fluids can vary not only due to internal effects within the system but also as a result of external influences. As such we need to monitor and measure the temperature of fluid power systems. This may be achieved with several devices. The following will be considered:

■ industrial thermometer
■ thermistor
■ thermocouple.

The industrial thermometer

The most common types of thermometer are those which employ the expansion of liquid within glass principle. The most popular liquids are mercury or alcohol and these types of device are accurate and reliable since they have no moving parts to develop faults.

Industrial thermometers use a metal bulb, often stainless steel, instead of glass. This is a robust thermometer which can be adopted for remote reading and indication by connecting the bulb to a Bourdon tube gauge via a suitable, flexible metal capillary tube as shown in Figure 11.12.

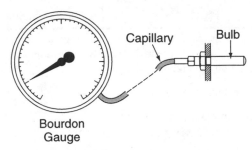

Figure 11.12 Industial thermometer with capillary tube and Bourdon Gauge

The thermistor

The thermistor is an electrical resistor that exhibits significant changes in electrical resistance with changes in temperature. Two types of thermistor are produced:

- the **PTC** (**positive temperature coefficient**) thermistor in which an *increase in temperature* is accompanied by an *increase in resistance*
- the **NTC** (**negative temperature coefficient**) thermistor in which an *increase in temperature* is accompanied by a *decrease in resistance*.

The property of change of resistance with change of temperature is used for temperature measurement purposes, and a typical arrangement is to connect the thermistor device into one arm of a Wheatstone bridge with a suitable display or recording device. Figure 11.13 shows an arrangement.

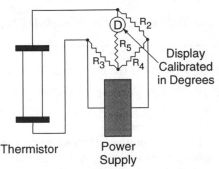

Figure 11.13 Thermistor/Wheatstone bridge temperature measuring system

The thermocouple

The thermocouple consists of two dissimilar electrical conductors joined together. When the two materials are part of a circuit with a measuring instrument there are two junctions and if there is a temperature difference between the two junctions then an electromotive force (emf) is set up. The size of this emf depends upon the difference in temperature and the materials involved.

Table 11.1 Illustration of typical thermocouple materials with their range and application

Type	Range	Application
Copper/constantan	−100°C to +400°C	General purpose
Iron/constantan	−100°C to +800°C	Domestic/commercial boiler temperature monitoring
Chrome/nickel-aluminium	−200°C to 1200°C	Cryogenic applications
Platinum/platinum-rhodium	0°C to 1600°C	Very high temperature applications

Note: Thermocouples enable the temperature to be measured over quite a small area, the area of junction between the two metals. Since they have a very small mass they can respond very rapidly to temperature changes i.e. they have good resolution and high sensitivity.

The sensitivities of two thermocouples are as follows:

- copper/constantan thermocouple: 0.03 mV/°C
- iron/constantan thermocouple: 0.05 mV/°C.

This shows that the iron/constantan thermocouple will give the larger voltage for a given change in temperature, and hence is the more sensitive.

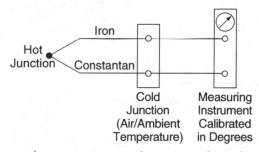

Figure 11.14 Thermocouple temperature measuring system using an iron/constantan device

11.5 Common terms used in measuring systems

It is clearly important to understand some of the terms used in measuring systems. Some important terms and their meanings are listed below.

- **Measurement.** The set of operations having the object of determining the **value of a quantity**.
- **Measurand.** The quantity being measured.
- **Error.** The difference between the **measured value** and a **standard value** obtained by **calibration**.
- **Accuracy.** The closeness of the measurement to the true value of the measurand.
- **Tolerance.** The range of inaccuracy which can be tolerated in measurement.
- **Resolution.** The smallest change in the measurand that can reasonably be measured
- **Sensitivity.** The relationship between the change in measurand and in the measured value.
- **Range.** The total range of values that an instrument/measuring system is capable of measuring

There are of course many more terms and **BS5233 1986 – Glossary of terms used in metrology** gives a full range of definitions of terms applied in measurement systems.

12 Troubleshooting and Maintenance

Aims

At the end of this chapter you should be able to:

1 Appreciate some of the causes and effects of malfunctions on pneumatic and hydraulic systems.
2 Be aware of the need for regular planned preventive maintenance checks and routines.

12.1 Introduction

Fault diagnosis involves a methodical and logical approach towards solving a problem. Faults within fluid power systems may occur due to:

- external failure, i.e. within the machine or process itself
- failure of the fluid power system.

In the case of machine or process failure, the action taken is largely dependent upon the complexity of the required repair work. Relatively simple problems can be resolved fairly rapidly, sometimes by process operators. More complex repairs often require service personnel to be called in.

With fluid power system failure, you should have obtained sufficient knowledge and experience of components and circuits in order to make an objective judgement of possible causes of failure.

12.2 Fault diagnosis – causes and effects of malfunctions

In any fault-finding exercise it is extremely useful to have the necessary documentation relating to the pneumatic and/or hydraulic system available. Such documentation should be supplied by the manufacturer/supplier of the fluid power system when it is delivered and/or installed. In general, the documentation for a fluid power system comprises:

- clear system layout diagrams with labelled valves and lines
- circuit diagrams
- list of components
- component data sheets
- displacement-step diagrams

- operating instructions
- installation and maintenance manuals
- list of spare parts for critical items.

In general, malfunctions of a system fall into the following categories:

1 Wear and tear on components and lines which can be accelerated by environmental influences:
   - quality of compressed air
   - relative motion of components
   - incorrect loading of components
   - incorrect maintenance or more usually the lack of maintenance
   - incorrect mounting and connection (i.e. signal lines are too long).
2 These conditions can lead to the following malfunctions or failures of the system:
   - seizure of units
   - breakages
   - leakages
   - pressure drop
   - incorrect switching.

Fault finding on fluid power systems

As a rule, a newly designed and installed fluid power system will run trouble-free for some time after initial adjustments have been carried out. Any instances of premature wear may not become noticeable until weeks or even months later. Normal wear may not become noticeable for years. Even then, faults or the effects of wear frequently do not become directly apparent, with the result that it is not easy to identify the defective component.

It is obviously not possible in this book to cover all the faults which may occur. The malfunctions described here are therefore those which occur frequently and which are difficult to localise in fluid power systems. Even more complex controls can be divided into smaller units and checked. In many cases the operator can eliminate the fault immediately, or at least identify the cause.

Malfunctions caused by undersized air supply in pneumatic systems

Frequently, sections of pneumatic systems are extended without enlarging the necessary air supply. Depending on the sequence and design of the plant section, malfunctions then occur not continuously but sporadically, with the result that fault finding is made increasingly difficult. Two possible effects that can arise are:

- the piston rod speed is not always correct, since the actuation of additional components can cause sudden pressure drops
- the force at the power cylinder drops for a short time during a pressure drop.

The same symptoms may occur as a result of the changes in orifice cross-sections caused by contamination or by leaks at connectors which have worked loose (a reduction in diameter of 20 per cent means a doubling of the pressure drop).

Malfunctions caused by condensate within pneumatic systems

It is essential to ensure that the compressed air fed into the network is free of condensate. Apart from the corrosive damage caused to surfaces by condensate which is, in many cases, extremely aggressive, there is the considerable danger of seizure of valve components if they need to be reset by spring force after being held in one switching position for a considerable time. Lubricants without additives have a tendency to emulsify and create resin or gumming. All close-tolerance sliding fits in valves are particularly susceptible to these resistances to movement.

Problems in hydraulic systems

Table 12.1 lists some common problems that occur in hydraulic systems.

Table 12.1 Problems in hydraulic systems

Problem	Possible cause
Excessive pump noise and fluctuating pressure	Pump cavitation caused by: clogged suction filter restricted suction line
Low flow under no load	Air leak in pump suction line, fittings, shaft seals, etc.; pump speed too high
Decreasing performance as pressure increases on one circuit	Faulty directional control valve; faulty cylinder or hydraulic motor
Decreasing performance as pressure increases on all circuits	Internal leakage in pump or main relief valve
Actuator fails to hold load	Faulty actuator, directional control valve or cross-line relief valve

The following effects and possible causes may occur in hydraulic systems:

1 **Dirty oil**
   - components not properly cleaned after servicing
   - air breathing not installed
   - tank not properly gasketted
   - pipelines not flushed when installed or after servicing
   - incorrect tank design without baffles to provide settlement traps
   - filters not replaced at correct intervals.
2 **Foaming oil**
   - return tank line not below fluid level; broken pipe
   - inadequate baffles in reservoir
   - fluid contamination
   - suction leak to pump allowing oil to aerate.
3 **Moisture in oil**
   - moisture in cans used to fill tank
   - water drain not fitted to lowest point in the tank
   - cold lines fastened directly against the hot tank causing condensation
   - cooling coils not below fluid level.
4 **Overheating of system**
   - water shut off or heat exchanger clogged
   - continuous operation at relief valve settings:
     – stalling under load etc.
     – fluid viscosity too high

- excessive slippage or internal leakage:
  - check stall leakage past pump, motors and cylinders
  - fluid viscosity too low
- tank size too small. The tank should be sized to approximately three times the capacity of the pump. Allowance should be made when using large bore cylinders for the amount of oil required for cylinder extension and retraction. This also applies where a large number of smaller cylinders are in the circuit
- inadequate baffling in the tank
- pipe or hose ID too small causing high fluid velocities to be achieved
- valves sized too small causing high fluid velocities to be achieved
- lack of air circulation around the oil tank
- system relief valve set too high
- ambient temperature around tank too high.

Malfunctions caused by contamination

Filters are generally fitted upstream in pneumatic and hydraulic systems. If, however, the supply lines for valves are not blown clear before they are connected, all the dirt particles produced by the connecting or welding process (sealing tape, welding beads, pipe scale, thread swarf and many other contaminates) can pass onto the valves.

In the case of pneumatic systems which have been in service for some time, an excessive proportion of condensate in compressed air may produce rust particles in cases where lines are fitted without corrosion protection. Contamination of pneumatic and hydraulic lines may produce the following effects:

- sticking or seizure of slide valve seats
- leaks in poppet valves
- blockage of flow control nozzles.

12.3 Maintenance of pneumatic systems

For pneumatic systems, the following regular maintenance routines are recommended:

- check the filter and service units – drain water regularly from traps and replenish and adjust lubricators where used
- discuss with the operators of the system any noted difference in performance or unusual events
- check for air leaks, crimped air-lines or physical damage
- check signal generators for wear or dirt
- check cylinder bearings and mountings.

Every day
Drain condensate from the filters if the air has a high water content, if no automatic condensate drainage has been provided. With large reservoirs, a water separator with automatic drain should be fitted as a general principle. Check the oil level in the compressed air lubricator and check the setting of the oil metering.

Every week
Check signal generators for possible deposits of dirt or swarf. Check the pressure gauge of the pressure regulators. Check that the lubricator is functioning correctly.

Every three months
Check the seals of the connectors for leaks. If necessary, re-tighten the connectors. Replace lines connected to moving parts. Check the exhaust ports of the valves for leaks. Clean filter cartridges with soapy water (do not use solvents) and blow them out with compressed air in the reverse of the normal flow direction.

Every six months
Check the rod bearings in the cylinders for wear and replace if necessary. Also replace the scraper and sealing rings.

12.4 Maintenance of hydraulic systems

For hydraulic systems the following regular maintenance routines are recommended:

- check oil level in all tanks
- check regulators and filters
- check pump drives
- check valve linkage for damaged parts
- check for external oil leaks
- check cylinder bearings and mountings
- discuss with the operators of the system any noted difference in performance or unusual events.

For both pneumatic and hydraulic systems the frequency of carrying out maintenance checks should be established on the basis of equipment and component manufacturers' recommendations together with the user's own field experience with the system.

12.5 A guide to the use of functional charts for fault finding

With the increased use of automation and thus of hydraulics and pneumatics, there is a growing need for a quick method of diagnosis of faults so as to minimise lost production. Functional charts provide one such means of doing this

The method of construction (refer to Figure 12.2)

Construct a chart with columns headed 'Item', 'Index', 'Rest', followed by one for each step of the sequence of operation * * * *, and a final one headed 'Stop'.

In the first column (headed 'Item') , list each piece of equipment in the circuit allowing one horizontal space for each condition the item can assume.

In the column headed 'Index' indicate the condition that each horizontal space represents, such as extended (+) or retracted (−), on or off, operative or inoperative, etc.

The column headed 'Rest' is used to show the condition of each item in the system *before it is set in motion*. Any changes in condition of each item are then recorded by the change of position of a line following a horizontal path across the chart. When indicating the movement of such items as cylinders, the movement is depicted by a sloping line (because the action takes time) but in the case of valves where the change is almost instantaneous, the line changes vertically.

Where a valve is not under operational conditions because it has no feed (such as during group changes in cascade circuits), the line is shown in dotted from so that a distinction can be made between a valve that will influence the action and a valve that has no bearing on the action at that moment. In this way it is possible to see at a glance which valves are of importance at the time of failure. If following the circuit diagram, it is necessary to remember the condition of each item as it changes during the sequence.

An additional aid is to thicken, colour or highlight the vertical lines where there is a change of group in a cascade circuit. Machines rarely breakdown in the 'start-up' condition, yet this is the state in which circuit diagrams are necessarily drawn.

For machines with a repetitive cycle these charts are ideal. For machines that have a programme which can be changed, it is necessary to make a chart for each part of the action (i.e. turret movement, cross slide movement), then use the part of the chart showing the action at the point of breakdown.

Method of use of a functional chart

Figures 12.1 and 12.2 show a system circuit and its functional chart respectively so that the method of construction is evident.

To make use of the functional chart, it is necessary to know at what point in the sequence of operations the system stopped. This information is usually supplied by the operator, or the product state. At this point of the chart, a rule may be placed vertically down the page and this will then indicate the valves etc. that changed at that moment in time. With this knowledge, it is not necessary to follow the whole sequence but just to check those valves indicated and the possible causes of their malfunction.

An extra point is that as a full line is used on the graph if the valve is being fed at that moment, and a dotted line where a different group is in use, or the valve is being bypassed, then the dotted line will automatically rule out any possibility of this unit being the cause of the breakdown provided the correct group, or supply, is under power at that moment.

It should be remembered of course that a unit can fail because it has not received a signal and this must also be checked.

Example of use of a functional cart

An example of the use of a functional chart would be if the system stopped at the point where cylinder C had completed its retraction back to the original position, but cylinder A had yet to retract for a complete cycle of operations.

Place a rule vertically down the chart on the line between the columns C– and A–, and it can be seen that DCVA (directional control valve for cylinder A) and item c - (roller-operated and spring-returned valve) have vertical changes in the graph, as also has valve 3 (3-port, 2-position, air-operated, spring-returned valve), but valve 3 has dotted lines at this point on the chart indicating that it has no feed and therefore has no effect on the circuit.

This means that it is only necessary to check that the other two items have been operated, or that they have a supply and a discharge.

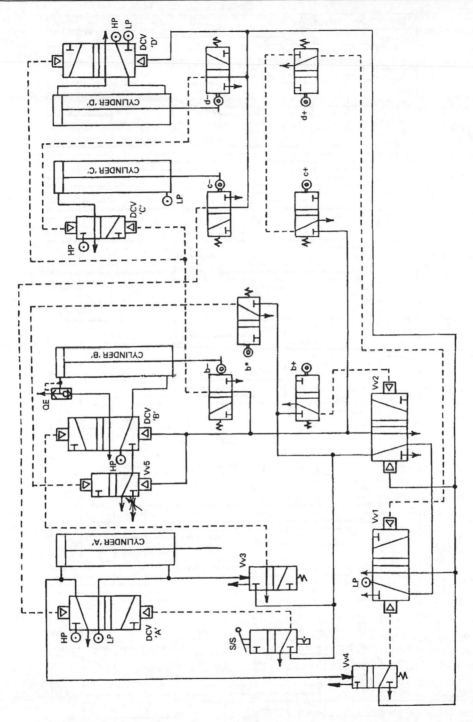

Figure 12.1 System circuit diagram. Sequence of operations: A+, B+, B–, (C+D+), D–, C–, A–, Stop/cycle
Key: DCV 'A' – Directional control valve; Vv1 & Vv2 – Group valves; S/S – Stop/start valve; HP – High pressure
DCV 'B' – Directional control valve; Vv3 & Vv4 – Diaphragm valves; Q/E – Quick exhaust valve; LP – Low pressure;
DCV 'C' – Directional control valve; Vv5 – Exhaust control valve (fitted with restrictor); b+b,* etc. – Trip valves;
DCV 'D' – Directional control valve

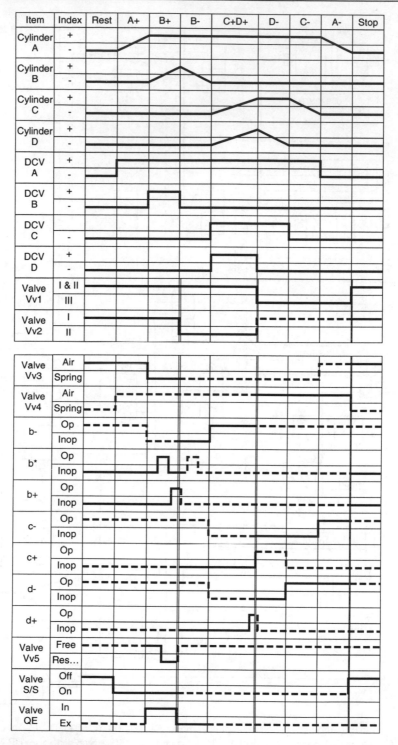

Figure 12.2 Functional chart diagram

Questions

Examine the circuit shown in Figure 12.3 and then attempt to answer questions 1–6 on page 118.

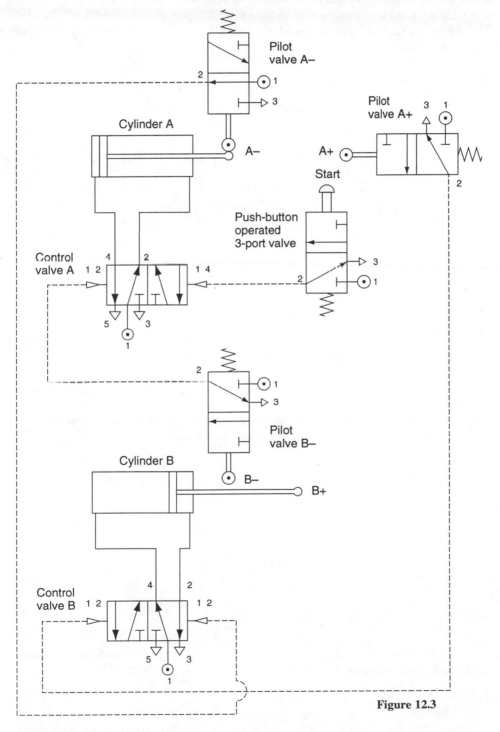

Figure 12.3

1 What fault could cause cylinder A to go positive when the 12 port of control valve A is pressurised?

2 What faults could cause unequal piston speeds?

3 The circuit runs through its sequence to B–. What should be the next movement ?

4 What faults could prevent this movement being achieved ?

5 Cylinder B has a leaking piston seal. Where will a tell-tale air leak occur ?

6 The spring fractures and fails to return pilot valve A+. What will be the likely consequence on circuit operation ?

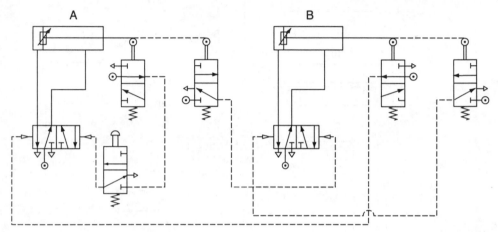

Figure 12.4

7 The circuit in Figure 12.4 was intended to run A+ B+ A– B–, but it has been incorrectly piped up. List the errors and re-draw the circuit for correct operation. Note: To assist you, refer to Section 9.6, pages 90 and 91.

8 If the air supply totally fails in the circuit shown in Figure 12.4, where will the cylinder come to rest ?

9 If the piston seal on cylinder A in Figure 12.4 is faulty, mark with an * the point from which the air will leak when the cylinder is in the position shown.

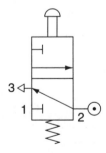

Figure 12.5

10 What is wrong with the connections shown in relation to the push button valve in Figure 12.5? State the effects that would be apparent in either switching position of the valve.

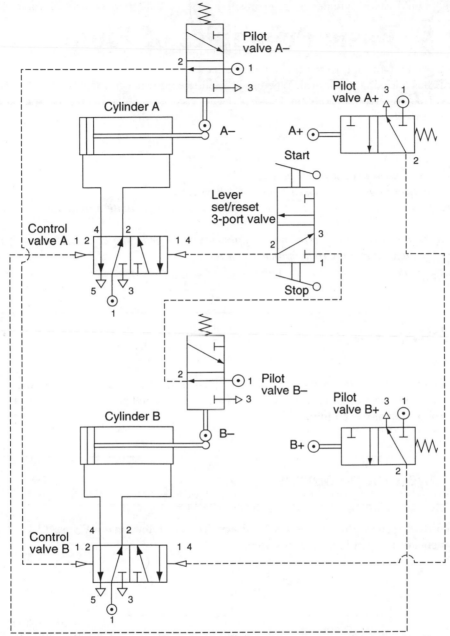

Stop/start control of the continuous cycle sequential control circuit A+, B+. A−. B−
(the cylinder A+ signal from pilot valve B− is inhibited by the stop/start valve)

Figure 12.6

11 In the circuit shown in Figure 12.6, imagine the circuit stops at A−, that is, will not go on to B−
What are the possible faults, beginning at the A− pilot valve?

13 Basic Principles of Fluid Power Control

Aims

At the end of this chapter you should be able to:
1 *Distinguish between open- and closed-loop control systems.*
2 *Recognise the various elements, components and signals that constitute a fluid power control system.*
3 *Appreciate the function of these elements, components and signals thereby understanding the principle of pneumatic and hydraulic process control.*
4 *Appreciate common terminology used in pneumatic and hydraulic process control.*

13.1 Introduction

In simple terms, a control system is merely a device which allows an operator to control the flow of energy to a machine or process in such a way as to achieve a desired performance. Control systems can be **open-loop** or **closed-loop**.

13.2 Open-loop control

Open-loop control systems do not possess feedback. The block diagram in Figure 13.1 shows the main elements of an open-loop control system.

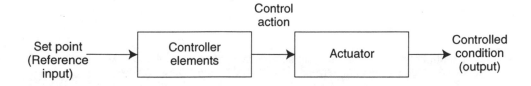

Figure 13.1 Open loop control block diagram

The principal features of open-loop control systems are:

- they are simple
- their accuracy is determined by the calibration of their elements
- they are not generally troubled with instability.

13.3 Closed-loop control

The block diagram in Figure 13.2 shows the main elements of a closed-loop control system.

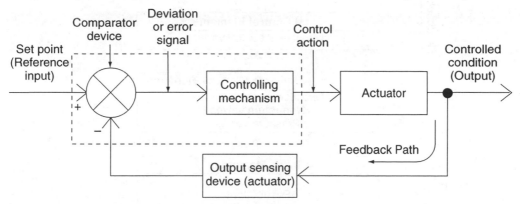

Figure 13.2 Closed loop (feedback) control block diagram

The principal features of closed-loop control systems are:

- they are highly accurate
- they are more complex
- non-linearities and distortion are greatly reduced
- they have wide bandwidth of operation
- they suffer from instability problems.

13.4 Signal flow and practical devices in fluid power control

The power section or work section consists of the actuator and the final control element. The control element receives control signals from the processor. The signal processor processes information sent from the signal input devices or sensors. The signal flow is from the energy source to the power section. See Figures 13.3 and 13.4.

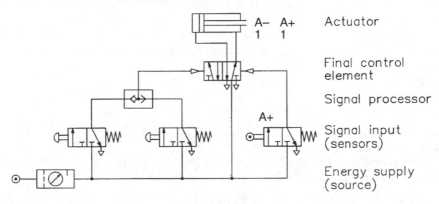

Figure 13.3 Circuit diagram of simple pneumatic control

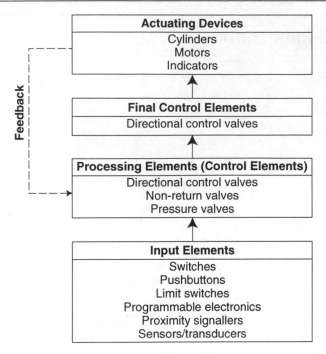

Figure 13.4 Signal flow in pneumatic and hydraulic control

13.5 Terminology

- **Set point (reference input).** The quantity or signal which is set or varied by some device or human agent external to and independent of the control system and which is intended to determine the value of the **controlled condition**.
- **Comparator.** The element which compares the monitored feedback signal with the externally applied **set point (reference input)**.
- **Deviation (error signal).** The difference between the measured value of the controlled condition and the **set point**.
- **Controller.** Those controlled elements and components required to generate the appropriate **control action** applied to the plant or process.
- **Control action.** The action (or actuating signal) that is taken by the control system in order to produce the required output. The control action may be applied independently or in combination:
  - **on-off**
  - **proportional**
  - **integral**
  - **derivative.**
- **Actuator.** The final control element in the control system.
- **Controlled condition.** The physical quantity or condition of the controlled body, process or machine for which it is the purpose of the system to control.

14 Emergency Shutdown and Safety Systems

Aims

At the end of this chapter you should be able to:

1 Understand the distinction between ultimate plant protection (emergency relief) devices, automatic and manual shut-down and process control safety functions.
2 Appreciate the various emergency shut-down actions in pneumatic and hydraulic circuits.
3 Understand the concept of 'fail safe' as applied to pneumatic and hydraulic circuits.

14.1 Introduction

For many years, process plant did not rely upon instrumentation and control systems for safety integrity. It was considered that the basic chemistry of the process itself contained by the mechanical integrity of vessels, pipework or buildings and limited by such apparatus as pressure relief valves, bursting discs, etc. were sufficient to ensure the safety of personnel and plant. Instrumentation, control and shut-down systems were provided purely for economic reasons, to prevent damage due to abnormalities, protect investment and to allow plant to resume profitable operation when normal conditions were re-established.

In recent years the trend has been to rely upon instrumented systems to ensure the safety of plant or more usually the safety of part of a process. This in turn has led to a structured approach towards ensuring safety (Figure 14.1).

Ultimate Protection
Containment
Pressure relief valves
Bursting discs   etc.
Automatic Shutdown
Pneumatic
Hydraulic
Electrical
Manual Shutdown
Pneumatic
Hydraulic
Electrical
Control System
Regulating process
Localised plant trips
Audible and visual alarms

Figure 14.1 Hierarchy of protection

Figures 14.2, 14.3 and 14.4 give simple illustrations of protection by containment, relief valve and automatic shut-down respectively.

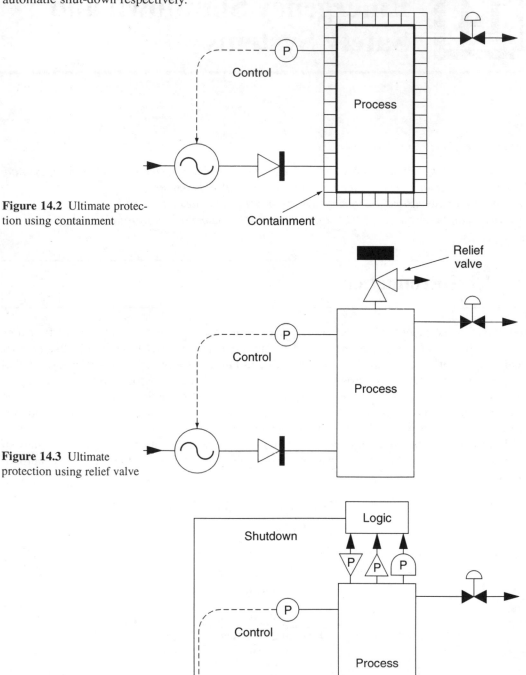

Figure 14.2 Ultimate protection using containment

Containment

Figure 14.3 Ultimate protection using relief valve

Figure 14.4 Protection using automatic shut-down

14.2 Emergency shut-down

An emergency condition could be said to exist when it becomes vital to stop the pneumatic or hydraulic sequence because of some unpredictable fault in the system in either the circuit itself, or perhaps in the process it controls.

Obviously, as more involved circuits are used, so the emergency stop circuits become more complex and additional equipment is needed to stop the circuit safely and quickly.

Perhaps the most obvious way of stopping a circuit in an emergency is to cut off (exhaust) the air supply, or in the case of hydraulics, to return fluid to tank. In certain situations this may not necessarily be the best thing to do because, for example, we may require control air or hydraulic fluid to be available in an emergency in order to drive or maintain a device in a safe or 'least hazardous' condition. The stop configuration of a cylinder in an emergency could be any one of four modes:

- to instroke
- to outstroke
- to stop
- to be exhausted – either instroke or outstroke.

In the foregoing arrangements circuits are drawn in the emergency condition. In Figure 14.5 the cylinder is normally controlled by the pilot circuit which operates the 5/2-way valve. In the event of an emergency the emergency valve cuts off the pilot signal and allows the spring in the 5/2-way valve to park the cylinder either in an instroked or outstroked position, depending on connections. This gives a predictable parking position which would not be possible if the emergency valve merely cuts off mains air or hydraulic supply.

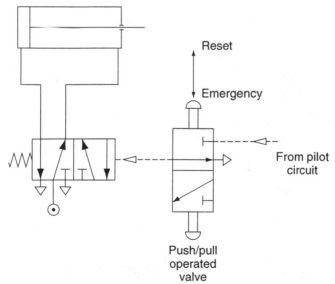

Figure 14.5 Cylinder positioned to safe position by indirect manual operation (using auto-return valve)

Both of the circuits in Figures 14.5 and 14.6 rely upon indirect action. Figure 14.6 applies to a very common circuit using a double pilot air-operated 5/2-way valve. In this case the emergency valve traps one pilot signal by operating valve X and at the same time breaks into the other pilot line and causes the 5/2-way valve to drive the cylinder to a safe parking position. The circuit can be reset by the remotely mounted reset valve. (It would be unwise to mount the emergency and reset valves in close proximity to one another.)

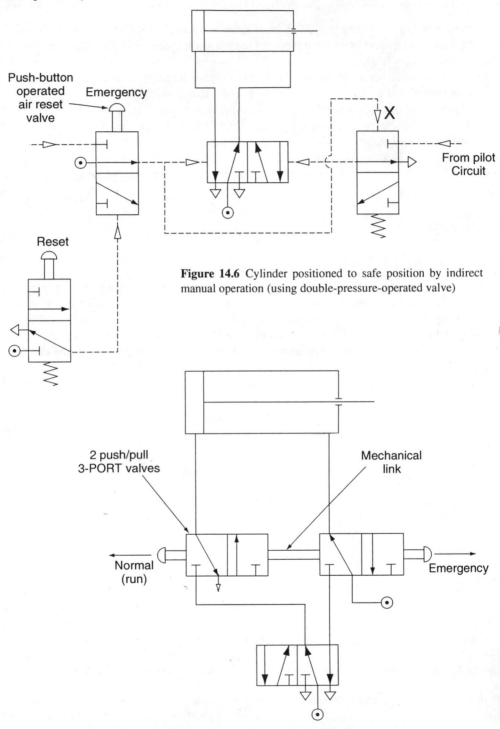

Figure 14.6 Cylinder positioned to safe position by indirect manual operation (using double-pressure-operated valve)

Figure 14.7 Cylinder positioned to safe position by direct manual operation

Where the safety of human life is concerned, a simple but effective direct-acting system is shown in Figure 14.7.

In all of the cases described so far, there is an obvious need for a protected supply of emergency air (or hydraulic fluid) to be provided, possibly from a reservoir or accumulator.

The requirement to stop a cylinder in mid-stroke is less easy to achieve accurately. Figure 14.8 shows one possibility, where a three-position control valve is used in which the springs centre the mid position.

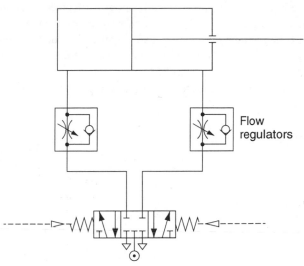

Figure 14.8 Fully sealed mid position valve normally spring loaded to mid position if pilot signals removed

In this case, the emergency valve would be arranged to remove the pilot signals so that the 5/3-way valve would centre and freeze the cylinder with fluid trapped on both sides of the piston. However, if the fluid was air, the piston would be unable to resist axial load because of the compressibility of the air.

Another possible arrangement for 'freezing' the cylinder piston in mid-stroke is shown in Figure 14.9. In this case if we were to remove the pilot air signals from the two 2/2 spring return valves then the cylinder would stop in mid-travel.

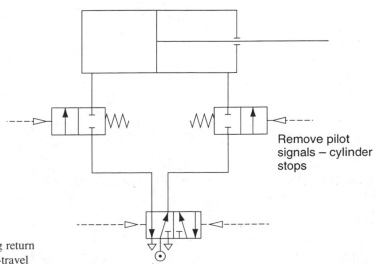

Remove pilot signals – cylinder stops

Figure 14.9 Use of 2/2 spring return valves to stop cylinder in mid-travel

The final emergency possibility, that of removing (exhausting) the air supply, (or returning hydraulic fluid to tank) would be quite simple to achieve, either by manual or piloted valve. The implications, however, must be considered. For example, the cylinders may end up parked in a hazardous condition, or on re-connection of the main air or hydraulic supply may accelerate sharply and cause damage.

In more complex multi-cylinder circuits it may be necessary to park the cylinders in a set sequence to prevent mechanical damage to the equipment they operate. This is shown in Figure 14.10 which refers to a three-group cascade circuit with emergency stop. When the emergency button is operated the pilot air, which is normally distributed by the mechanical trip valves, is cancelled. At the same time cylinder control valves and group change-over valves are reset to make cylinders D and E instroke. (Note the shuttle valves are used to achieve this.) When D and E have instroked, valves 1 and 2 are tripped and signal the control valves for A and C to instroke.

When the reset button is operated the pilot air is reinstated and the circuit will begin to operate in its normal sequence.

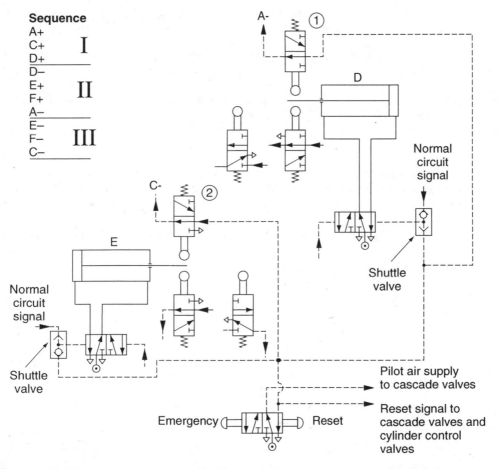

Figure 14.10 Three group cascade circuit with emergency stop to ensure that all cylinder pistons return to instroke position. Note: only part of the circuit is shown to illustrate the principle (circuit is shown in reset position)

14.3 Fail safe

A fail safe circuit is normally designed with the intention that the system shall fail to the least hazardous condition and in a predetermined configuration in the event of the air or hydraulic supply failing.

To achieve this, it is normal to assume a separate protected air or hydraulic supply for the fail safe circuit. Also a pressure-sensitive spring-operated valve is needed to sense once air or hydraulic supply falls below the safety threshold value.

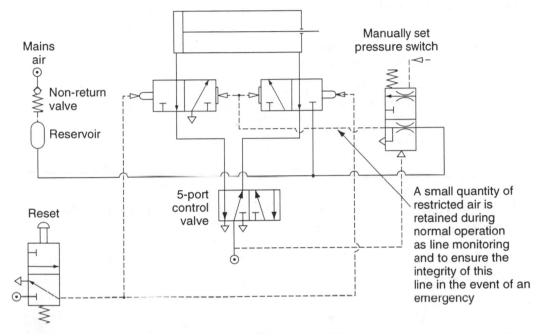

Figure 14.11 A fail-safe circuit for double-acting cylinder control. Note: the circuit is shown in normal operating condition

In Figure 14.11, the manually set pressure switch detects a fall in main air pressure and switches the pair of 3/2 priority valves, using protected air from the reservoir. This protected air is also used to next drive the cylinder negative to a safe parking position. If the reset valve is pressed before normal air supply is reinstated, the priority valves remain switched in the emergency position and ensure that the cylinder remains instroked until main air is restored.

Exercise

You may wish to consider a fail safe circuit requiring the cylinder to stop (freeze) in mid-stroke. Sketch a circuit, using a pressure sensitive switch in order to achieve this effect.

15 Health and Safety at Work

Aims

At the end of this chapter you should be able to:

1 Appreciate the organisational structure under the Health and Safety at Work Act.
2 Be aware of the employer's responsibilities under the Act.
3 Be aware of the employee's responsibilities under the Act, in particular:

   (a) the need to cooperate with the employer in respect to health and safety matters
   (b) the need for the employee to take all reasonable care and to ensure the safety of him/herself and others in respect to matters within his/her control.

4 Be aware of recent regulations applicable to compressed air systems.

15.1 Organisation under the Health and Safety at Work Act 1974

The basic structure of the UK Government's Health and Safety organisations is depicted in Figure 15.1.

Figure 15.1 Hierarchy of health and safety bodies within the UK

The principal role of the HSE may be summarised as:

1 produce safety-related legislation, guidance and codes of practice
2 enforce legislation through improvement and prohibition notices etc.
3 carry out inspections.

15.2 Employer's responsibilities

Under Section 2 of the Health and Safety at Work Act 1974 duties are placed upon an employer in respect to the following:

■ **Provide and maintain safe plant and systems of work (HSWA Section 2 (2) (a)).**
■ **Make arrangements for ensuring safe use and handling of articles and substances (HSWA Section 2 (2) (b)).**
■ **Provide adequate information, instruction, training and supervision (HSWA Section 2 (2) (c)).**
■ **Maintain a safe place of work (HSWA Section 2 (2) (d)).**
■ **Provide and maintain a safe working environment (HSWA Section 2 (2) (e)).**

All of the above are subject to the qualification: **so far as is reasonably practicable.**

It is well established in law that for a requirement to be made **so far as is reasonably practicable**, an assessment has to be made of the magnitude of the particular risk and the expense of eliminating or minimising it. The greater the risk, the less weight given by the courts to the cost of preventing it.

Under Section 2(3) of the Health and Safety at Work Act it shall be the duty of every employer to prepare and revise a written statement of his/her general policy with respect to the Health and Safety at work of his/her employees and the organisation and arrangements for carrying out that policy. This generally requires an employer with five or more employees to provide a written **safety policy document**.

15.3 Employee's responsibilities

Under Section 7 of the Health and Safety at Work Act it shall be the general duty of the employee to:

■ **safeguard him/herself and others**
■ **cooperate with the employer in respect to health and safety matters**
■ **use equipment and facilities provided to ensure safety and health at work.**

15.4 Pressure systems and transportable gas containers regulations 1989 (SI 1989 No 2169)

These regulations, which are applicable to compressed air systems operating at 0.5 bar above atmospheric pressure, came into force from 1 July 1990. They were introduced over a four-year transitional period. A list of individual regulations together with the dates when they came into force are given on the next page.

Effective from 1 July 1990

Regulation 1   Citation and commencement
Regulation 2   Interpretation
Regulation 3   Application and duties
Regulation 5   Provision of information and marking
Regulation 6   Installation
Regulation 7   Safe operating limits
Regulation 13   Keeping of records etc.
Regulation 14   Application
Regulation 15   Precautions to prevent pressurisation
Regulation 23   Defence
Regulation 24   Power to grant exemptions
Regulation 27   Transitional provisions

Effective from 1 January 1991

Regulation 4   Design, construction, repair and modification (in so far as it relates to transportable gas containers)
Regulation 16   Design standards, approval and certification
Regulation 17   Filling of containers
Regulation 18   Examination of containers
Regulation 19   Modification of containers
Regulation 20   Repair work
Regulation 21   Re-rating
Regulation 22   Records
Regulation 26   Repeals, revocations and modifications (in so far as it relates to transportable gas containers)

Effective from 1 July 1994

Regulation 4   Design, construction, repair and modification (in so far as it relates to pressure systems)
Regulation 8   Written scheme of examination
Regulation 9   Examination in accordance with the written scheme
Regulation 10   Action in case of imminent danger
Regulation 11   Operation
Regulation 12   Maintenance
Regulation 26   Repeals, revocations and modifications (in so far as it relates to pressure systems)

General aspects of the regulations

The overall intention of the regulations is to prevent the risk of serious injury from stored energy as the result of the failure of a pressure system or part of it.

The regulations are concerned with steam and *gases under pressure in excess of 0.5 bar and fluids which are kept under pressure and become gases on release to the atmosphere*. It should be noted that the regulations do not concern themselves with completely hydraulic systems, where the stored energy is relatively low.

The regulations impose duties on the **owner** and **user** of installed and mobile pressure systems and gas cylinders (termed 'transportable gas containers' within the regulations).

Competent persons are required to carry out the periodic inspections of pressure systems and to draw up an initial written scheme of examination. Within the context of these regulations, the 'competent person' refers not to the individual employee who carries out duties under the regulations but only to the body which employs him/her.

The following bodies may provide competent person services:

- a user company with its own in-house inspection department
- an inspection organisation providing such services to clients
- a partnership of individuals
- a self-employed person.

HSE documentation

As is usual with regulations of this nature, the HSE has published a selection of *Guidance and Approved Code of Practice (ACOP)* documents as follows:

- HS (R) 30 – A guide to the Pressure Systems and Transportable Gas Containers Regulations
- ACOP – Safety of Pressure Systems
- ACOP – Safety of Transportable Gas Containers.

In addition, the ACOPs list relevant *HSE Guidance Notes* which form a second tier of guidance on particular areas of work related to pressure systems plant and equipment.

The above documents are intended to offer assistance to practitioners by giving them the HSE's interpretation of the regulations. The documents will be of particular use to technical management, engineers and technicians.

Categories of pressure systems

For the purpose of the regulations, pressure systems are divided into three categories:

Minor systems

These include those containing steam, pressurised hot water, compressed air, inert gases or fluorocarbon refrigerants. Pressure must be below 20 bar and the largest vessel should not exceed 200 000 bar litres. (This will include most compressed air systems.) Note that systems of less than 250 bar litres are exempt. Temperatures are between –20°C and 250°C.

Intermediate systems

The majority of storage and process systems, including pipelines, fall into this category, unless they are minor or major systems.

Major systems

These are very large and complex systems, including steam generation in excess of 10 MW per generator and any pressure storage system where the largest pressure vessel is more than 1 000 000 bar litres.

Note: The term '**bar litres**' may not be entirely familiar to some readers. In this respect, Appendix 5 illustrates the concept with some simple calculations.

Layout of the regulations

There are 27 individual regulations divided into four parts with an additional six schedules relating to the regulations. The following is a brief review summary of each of the main technical provisions of the regulations:

Regulation 1 – citation and commencement

This is the citation and commencement, which governs the dates on which individual regulations became effective.

Regulation 2 – interpretation

This defines terms which are used within the regulations. Of particular importance are:

Competent person

As previously mentioned, a competent person is not considered to be an individual employee who carries out duties under the regulations. It refers to the body which employs him/her. However, a competent person may be an individual if that person is self-employed and provides competent person services.

Danger

This emphasises that the regulations are intended to address the dangers caused by the uncontrolled release of stored energy, rather than other dangers which may be covered by existing legislation. Note that steam is covered by these regulations.

Examination

In this context the format of the inspection to be carried out under the written scheme of examination:

Installed system and mobile system

For a fixed system the **user** is responsible for compliance with Regulations 7–13 and for mobile systems the **owner** has responsibility. But note that in certain cases, the owner of a leased fixed system may retain responsibility for its safe use.

Pipeline

This means the pipeline(s) used for the transmission of relevant fluid across the boundaries of premises, together with associated valves, pumps, compressors, etc., but only pipelines for gases and liquefied gases are covered as a pressure system (other regulations do, however, apply).

Pipework

Not to be confused with pipelines. In the pressure systems context the pipework includes valves, compressors, cylinders etc. In other words, almost anything except pressure vessels and protective devices.

Pressure system

Three types of system are defined:
1 a system comprising a pressure vessel, associated pipework and protective devices. The pressure vessel defines the system
2 pipework and protective devices to which a gas cylinder may be attached. Pressure must be over 0.5 bar
3 a pipeline with its protective devices.

Protective devices

A protective control or measuring system which is essential to prevent a dangerous situation from arising.

Relevant fluid

There are three types:
1 steam
2 fluid or mixture of fluids which is at a pressure greater than 0.5 bar and which is:
   (i) a gas
   (ii) a liquid having a vapour pressure greater than 0.5 bar when in equilibrium with its vapour at either the actual temperature of the liquid or 17.5°C, or
3 a gas dissolved in a solvent, e.g. acetylene.

Safe operating limits

Operating limits (including a suitable margin of safety) beyond which system failure is likely to occur.

Scheme of examination

A suitable written scheme drawn up or certified by a competent person for the examination at appropriate intervals of most pressure vessels and all safety devices, and any pipework which is potentially dangerous. The user may seek the advice of any competent person when deciding what vessels and parts of the pipework need to be included in the scheme.

Regulation 8 deals specifically with the written scheme of examination and **Regulation 9** refers to the examination in accordance with the written scheme.

Transportable Gas Container

Robust transportable gas cylinders having a capacity between 0.5 and 3000 litres if refillable and 1.4 and 5 litres if intended to be discarded.

User

The person or corporate entity normally responsible for the safe operation of a fixed pressure system. Note that with some leased but fixed equipment the owner may retain responsibility (see below).

Regulation 3 – application and duties

This regulation makes it clear that the regulations are intended to secure the safety of people at work, both self-employed as well as employed. In certain cases the owner of a leased fixed system may retain responsibility for its safe use.

Regulation 4 – design, construction repair and modification

This regulation is concerned with the initial integrity of plant. It applies to newly fabricated plant and second hand plant supplied after the regulation came into effect.

The onus is placed on designers and fabricators to consider at the manufacturing stage both the purpose of the plant and how it will comply with the regulations. In addition, any modification or repair shall be such that it does not give rise to danger or otherwise impair the operation of any protective device or inspection facility.

See pages 13–14 of *HSE Guidance* and pages 9–4 of *ACOP*, which give specific guidance for the design and repair of steam, hot water and compressed air systems.

Regulation 5 – provision of information and marking

Requires designers and manufacturers to supply written information concerning design, construction, examination, operation and maintenance of equipment. The employer of a person who modifies a system has a duty to supply relevant details.

Pressure vessels shall be marked as in Schedule IV. No one shall tamper with these markings.

Regulation 6 – installation

Puts specific onus on installers to ensure that equipment is fitted correctly, gives no rise to danger and does not impair the action of any protective device or inspection facility.

Regulation 7 – safe operating limits

Users or owners shall not operate or allow to be operated systems unless he/she has established the safe operating limits.

The owner of a mobile system shall supply the user with a written statement of safe operating limits of the system or mark the equipment durably and legibly in a clearly visible place.

Regulation 8 – written scheme of examination

This is one of the most wide ranging regulations which puts new responsibilities on owners and users and requires a suitable written scheme for the periodic examination of those parts of pressure systems in which a defect may give rise to danger. The scheme shall be drawn up or certified as being suitable by a **competent person**:

- see pages 17–19 *HSE Guidance* and pages 18–21 in *ACOP.*
- see pages 2–3 of *AOTC Guidance on Examinations of Compressed Air Systems* for details of content of written scheme of examination.

Regulation 9 – examination in accordance with the written scheme

This regulation deals with the routine examinations carried out under the written scheme. The user or owner is not responsible for the quality of the examination, although he/she is responsible for ensuring that inspections are carried out at the appropriate times and that the system is not operated outside of its inspection period.

Mobile systems must have their date of next inspection clearly marked (but not engraved). Regulation 9.7 allows for the postponement of the next examination in certain circumstances. In all cases the user of fixed systems is responsible for preparing it for examination.

See AOTC booklets for details of suggested inspection frequencies, and HSE Guidance pages 19–22 and ACOP pages 21–25.

Regulation 10 – action in case of imminent danger

This regulation has application in the event of imminent danger where the competent person detects serious defects which require immediate attention.

The competent person shall issue a written report to the owner or user, specifying the nature of the defect and the immediate action required. The owner or user shall not operate the system until repairs have been undertaken. Additionally, the competent person shall notify the HSE within 14 days of the completion of the inspection as to the nature of the defects.

- See HSE Guidance pages 22–23 and ACOP page 25.

Regulation 11 – operation

The user or owner shall provide for any person operating the system adequate and suitable instructions for:

- safe operation of the system
- action to be taken in an emergency.

Users shall also ensure that the system is not operated except in accordance with the instructions provided. The implications here are that training, verbal instructions, written notices, signs, posters and aide-memoires will all be forms of operating instructions. Flow diagrams may also be helpful.

Regulation 12 – maintenance

The user or owner shall ensure that the system is maintained in good repair, so as to prevent danger. Note that **safety** is the reason for maintenance, rather than cost or improved efficiency. Consideration must be given to an appropriate maintenance system, record of work done, hours run, manufacturer's servicing data, etc.

■ See ACOP page 28.

Regulation 13 – keeping of records

Owners and users shall keep defined documents relating to pressure systems as follows:

■ last inspection report
■ any relevant previous ones
■ documents supplied under Regulation 5
■ documents to be passed on to new owner
■ See page 24 of *HSE Guidance*.

Regulation 14 – applications

This part of the regulations applies to a vessel which has a permanent outlet to atmosphere which if blocked could lead to over-pressurisation.

Regulation 15 – precautions to prevent pressurisation

The owner has a duty to ensure that outlets referred to in Regulation 14 are maintained in an unblocked state. Regulations 14 and 15 extend provisions relating to steam containers to all pressure systems.

Schedule I

This deals with the coming into force of the regulations and arrangements made during the transitional period.

Schedule II

Covers exceptions to the regulations of which there are 25 covering a wide variety from slurry tankers to fire extinguishers, with additional partial exemptions for several other cases.

Schedule III

The modification of owner-user duties where equipment is supplied under a lease or hire is explained in this schedule.

Schedule IV – marking of pressure vessels

From Regulation 5, pressure vessels must be marked with:

1 manufacturer's name
2 serial number
3 date of manufacture
4 standard to which built
5 maximum design pressure
6 minimum design pressure if less than atmosphere
7 design temperature.

Schedule V – fees for approvals

Deals with HSE fees for the approval of a quality assurance scheme, design specification or person(s).

Schedule VI – repeals, revocations and modifications

This final schedule details the legislation to be amended as the new regulations take effect, and is linked to Regulation 26 of the same name.

Questions

1 The regulations are intended to prevent the risk of serious injury caused by the uncontrolled release of what?
2 Within the scope of the regulations, who is responsible for the safe operation of a mobile system?
3 In a minor system, what is the maximum value in bar litres of the largest vessel?
4 An exempt vessel will be less than how many bar litres?
5 State two requirements of a 'competent person'.
6 Define the difference between pipeline and pipework.
7 Check Regulation 5 and list seven items of information which should be marked on an air receiver.
8 State two features of the written scheme of examination.
9 Who is responsible for ensuring the safe operation of a fixed pressure system?
10 Regulation 12 is concerned with maintenance, but with what in mind?

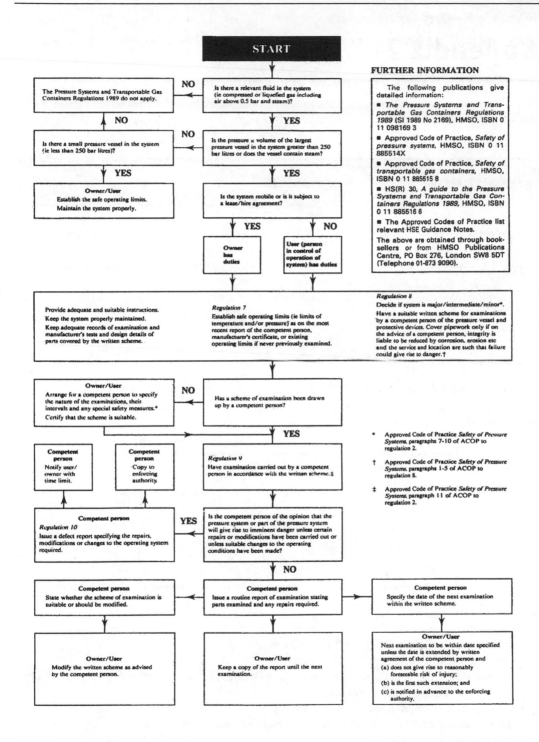

Figure 15.2 Owner's and users' duties under the regulations for existing plant (excluding gas cylinders, pipelines and open vented vessels)

Appendix 1

Answers to revision questions

Chapter 2

1 Nitrogen – 78% volume
   Oxygen – 21% volume
2 (a) force – newtons
   (b) area – square metres
   (c) volume – cubic metres
   (d) flow rate – cubic metres per second
   (e) pressure – pascal or more usually bar
3 1 bar
4 Approximately 1.75 m^3 (Boyle's law)

Chapter 3

1

Absolute pressure (P_{ab}) (bar)	Gauge pressure (P_g) (bar)
6	**5**
3.5	2.5
17	**16**
9.9	8.9

2 1 cubic metre and 4 bottles remaining under pressure.
3 **Zone 1 hazardous operation – advantage**
   Compressed air and hydraulics offer minimal risk of explosion or fire and no expensive protection against explosion is required.
   Fast working speed of operation – advantage
   Compressed air and hydraulics are very fast working media thereby enabling very high working speeds to be obtained.
   Low ambient noise level – disadvantage in pneumatics
   Exhaust air is loud. However, with suitable sound absorption material and silencers the noise level may be reduced to acceptable limits. This is not a problem with hydraulics.

Chapter 4

1

One-way adjustable flow control valve (flow regulator)		Filter with water trap automatically drained	
Lubricator		Silencer	
Air service unit		Shut-off valve (e.g. gate valve) (simplified)	
Actuation by lever		Four-port valve two positions	
Single-acting cylinder returned by spring		Shuttle valve	
Solenoid operation		Pilot control line	- - - - - - - - - - -
Non-return valve e.g. check valve		Two-port valve	
Double-acting non-cushioned cylinder		Double-acting cylinder with double adjustable cushioning	
Drain without pipe connection		3/2-way directional control valve	
Spring operation		5/3-way directional control valve mid position closed	

2

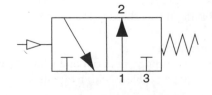

(a)

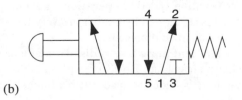

(b)

3   (i)   Air service unit comprising filter, regulator, gauge and lubricator.
(ii)   3/2-way valve with pressure, exhaust and signal ports. Double direct pneumatic actuation.
(iii) 3/3-way valve with closed neutral position and pressure, exhaust and signal ports. Double
solenoid and pilot operation with manual override.
(iv) Hydraulic pump.
(v)   Accumulator or air receiver.
(vi) 5/3-way valve, neutral position working lines vented. Pressure exhaust and signal ports.
Double direct pneumatic actuation.

4

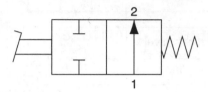

Chapter 15

1   Stored energy
2   Owner
3   200 000 bar litres
4   250 bar litres
5   Two from this list:
■ practical knowledge
■ theoretical knowledge
■ adequately qualified
■ actual experience of the type of system being examined
6   **Pipelines** are used for the transmission of fluids across boundaries of premises whereas
pipework means a pipe or system of pipes together with associated valves, pumps, compressors and other pressure-containing components excluding pressure vessels and protective
devices.
7   Manufacturer's name
serial number
date of manufacture
maximum design pressure
design temperature
standard to which built
minimum design pressure if less than atmosphere
8   Two from this list:
■ identification of equipment to be examined in the scheme
■ inspection frequency
■ competent person to draw up or certify the written scheme.
9   User
10   Safety of operation

Appendix 2

1   Define the term 'maintained signal' as used in pneumatic systems.   (6 marks)
2   (a) Define the term 'two-stage compression'.   (2 marks)
    (b) State two reasons for preferring multi-stage compression to single-stage
        compression for high pressure.   (4 marks)
3   Name the four components that make up a service unit.   (6 marks)
4   (a) State the function of a back-up ring on a piston in a pneumatic actuator
        with reference to an O-ring seal.   (2 marks)
    (b) Show, by means of a neat sketch, the action of a cup seal under pressure.   (4 marks)
5   Show, by means of a neat sketch, the BS 2917 symbol for a double-acting
    cylinder with variable cushioning in both directions.   (6 marks)
6   (a) Name two types of displacement compressor.   (2 marks)
    (b) State how a displacement compressor differs from a dynamic compressor.   (4 marks)

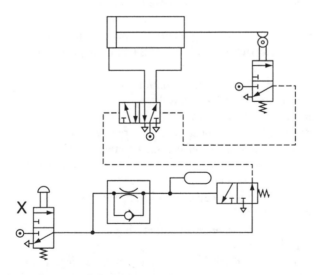

Figure A2.1

7   Referring to Figure A2.1, state the effect on the piston:
    (a) Of giving a momentary signal on valve X   (2 marks)
    (b) If valve X is held open.   (2 marks)

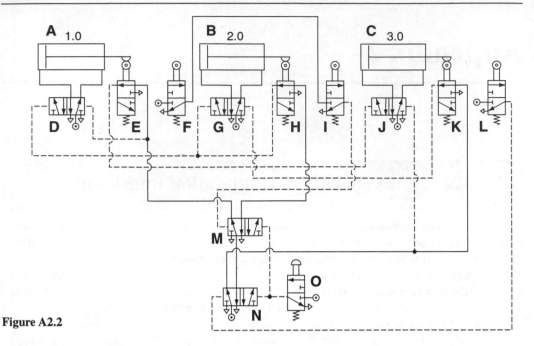

Figure A2.2

Questions 8, 9 and 10 refer to Figure A2.2

8   Determine the sequence of operation of the circuit and divide the
sequence into cascade groups.   (6 marks)

9   State the effect of holding valve O open.   (6 marks)

10 (a) Indicate, using the component letters given, which valves are
interlocking valves.   (2 marks)
(b) State the reason for using interlocking valves in this circuit.   (4 marks)

11 Draw, to BS 2917, the pneumatic symbol for a 3/2 manually operated,
normally open valve.   (6 marks)

12 State one reason for preferring pneumatic actuation to:
(a) Electrical actuation in an explosive atmosphere   (2 marks)
(b) Hydraulic actuation in the food industry   (2 marks)
(c) Mechanical actuation where a number of levers are necessary.   (2 marks)

13 (a) State three functions of an air receiver in a pneumatic system.   (3 marks)
(b) (i) name the type of air compressor which does not necessarily need an
air receiver.
(ii) state the reason why an air receiver is not necessary with this type
of compressor.   (3 marks)

14 (a) State the effect upon piston movement of 'metering in' air to a
pneumatic cylinder.   (2 marks)
(b) List four factors affecting the speed at which a pneumatic piston moves.   (4 marks)

15 Figure A2.3 shows a multi-position linear actuator in position 1. If position 1
is symbolised by A– B–, state the symbols for positions:
(a) 2   (2 marks)
(b) 3   (2 marks)
(c) 4   (2 marks)

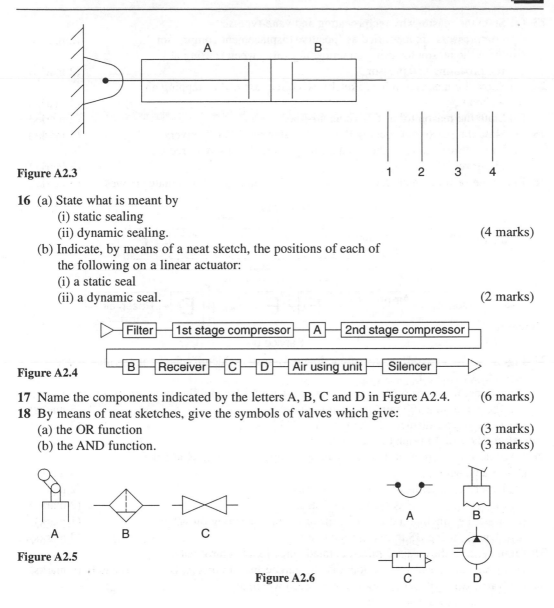

Figure A2.3   1   2   3   4

16 (a) State what is meant by
       (i) static sealing
       (ii) dynamic sealing.   (4 marks)
   (b) Indicate, by means of a neat sketch, the positions of each of
       the following on a linear actuator:
       (i) a static seal
       (ii) a dynamic seal.   (2 marks)

Figure A2.4

17 Name the components indicated by the letters A, B, C and D in Figure A2.4.   (6 marks)
18 By means of neat sketches, give the symbols of valves which give:
   (a) the OR function   (3 marks)
   (b) the AND function.   (3 marks)

Figure A2.5

Figure A2.6

19 Name each component symbolised in Figure A2.5.   (6 marks)
20 Figure A2.6 shows symbols from BS 2917. Identify the component
   represented by each symbol.   (6 marks)
21 State, with reference to pneumatic actuation compared with electrical
   actuation, two advantages of using pneumatic actuation for:
   (a) Linear motion   (3 marks)
   (b) Rotary motion   (3 marks)
22 (a) State the effect on moisture content as air is cooled.   (2 marks)
   (b) State two reasons why compressed air may need to be heated.   (2 marks)
   (c) Name two contaminants removed from compressed air by filtration.   (2 marks)

23 (a) State the reason why reciprocating and vane-type air compressors are identified as 'positive displacement compressors'. (2 marks)

(b) State the reason for using an unloading valve when starting a reciprocating compressor. (2 marks)

24 (a) Show, by means of a neat sketch, the correct method of tapping an air main. (4 marks)

(b) State the reason for a 'fall' in an air-line. (2 marks)

25 (a) State the reason for locking the safety valve on an air receiver. (3 marks)

(b) Name three fittings, other than a safety valve, legally required on all air receivers. (3 marks)

26 List three factors which determine the speed at which a linear actuator moves. (6 marks)

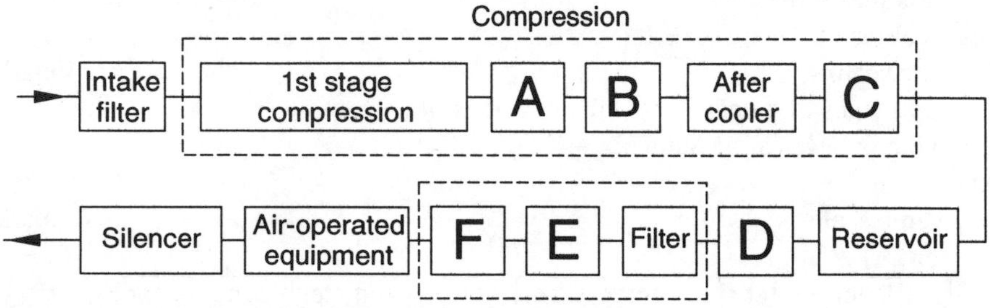

Figure A2.7   Service unit

27 Figure A2.7 shows a diagram of an air supply system. Identify the components lettered A, B, C, D, E and F.

28 Determine, using the nomogram given as Figure A2.8 and a line pressure of 4 bar:

(a) The force on a 100 mm diameter cylinder (3 marks)

(b) The cylinder diameter if the force is 2000 N (3 marks)

Questions 29 and 30 refer to Figure A2.9.

29 (a) State why there would be delay between operating valve X and the piston advancing. (4 marks)

(b) Describe, in terms of its functions, valve V. (2 marks)

30 (a) State why valve H is incorrectly fitted. (2 marks)

(b) State the function indicated by the arrow on the spring on valve Y. (1 mark)

(c) Identify, using BS terms, component Y. (3 marks)

31 Figure A2.10 shows a 5/2 pilot-operated, directional control valve. Using the numerical system, identify the numbering of ports A, B, C, D, E and F. (6 marks)

32 (a) Name two cylinder components between which a:

(i) static seal would be fitted

(ii) dynamic seal would be fitted. (4 marks)

(b) Illustrate, by means of a neat sketch, how an O-ring seal may extrude under pressure. (2 marks)

33 Draw to BS 2917 the symbols representing each of the following

(a) Through rod cylinder (2 marks)

(b) Differential cylinder (2 marks)

(c) Cylinder with internally cushioned piston. (2 marks)

34 Draw to BS 2917 the symbols representing each of the following:

(a) A variable speed pneumatic motor (2 marks)

(b) A silencer (1 mark)

(c) A service unit. (3 marks)

35 List four factors which affect the flow of air through a pneumatic system. (6 marks)

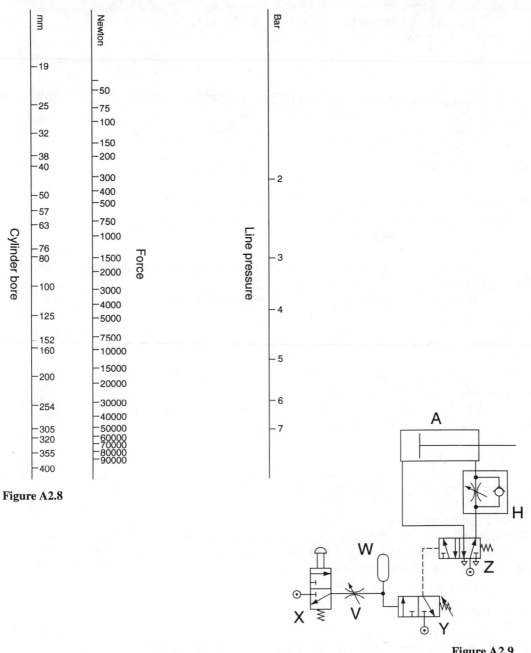

Figure A2.8

Figure A2.9

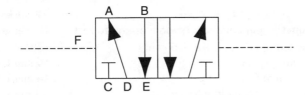

Figure A2.10

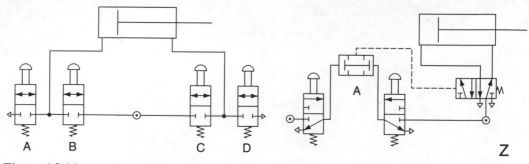

Figure A2.11   **Figure A2.12**

36 Figure A2.11 shows a circuit operated by means of 2/2 valves.
   (a) State which valves need to be operated to:
      (i) advance the piston
      (ii) retract the piston.   (4 marks)
   (b) State the effect on the piston of opening valves B and C at the same time   (2 marks)
37 (a) Identify valve A in Figure A2.12.   (2 marks)
   (b) State one possible reason for including valve A in a circuit.   (4 marks)

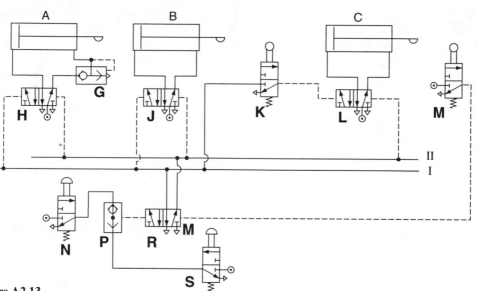

Figure A2.13

Questions 38, 39 and 40 refer to Figure A2.13
38 (a) State the sequence in which the circuit will reciprocate.   (4 marks)
   (b) State the number of cycles the circuit will complete before coming to rest. (2 marks)
39 (a) Identify and state the function of valve P.   (2 marks)
   (b) State which piston positions are not proven.   (4 marks)
40 (a) Identify and state the function of valve G.   (4 marks)
   (b) Name the system used in Figure A2.13 to avoid maintained
      (locked-in) signals.   (2 marks)

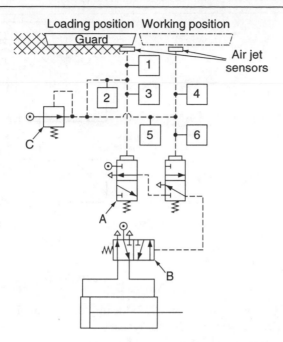

Figure A2.14

41 Figure A2.14 shows a pneumatic circuit where the operation of the cylinder is initiated by the closing of the guard. Air jet sensors are used to detect the guard movement.
(a) Using British Standard terms, describe each of the valves A, B and C.　(9 marks)
(b) Identify the positions in which restrictors would be placed using the numbered boxes in Figure A2.14.　(3 marks)

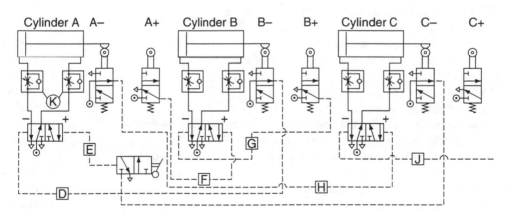

Figure A2.15

42 For the pneumatic circuit shown in Figure A2.15 state:
(a) The sequence in which the cylinders will operate assuming all maintained (trapped) signals have been overcome　(4 marks)
(b) The maintained signals present using the letters E to J　(4 marks)
(c) The effect of mounting the two valves marked K so that air is metered into the cylinder.　(4 marks)

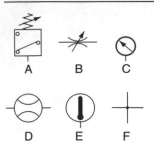

Figure A2.16

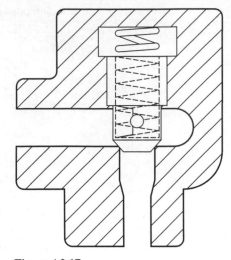

Figure A2.17

43 Figure A2.16 shows symbols drawn to BS 2917. Identify each symbol.   (6 marks)
44 Figure A2.17 shows a valve used in hydraulic systems.
    (a) State the common name of the valve.   (2 marks)
    (b) Draw the symbol, in accordance with BS 2917, to represent the valve.   (2 marks)
    (c) State its function.   (2 marks)
45 (a) State what is meant by:
      (i) static seal
      (ii) dynamic seal.   (3 marks)
    (b) Name one application of each of the following:
      (i) an O-ring seal as a static seal
      (ii) a U-ring seal as a dynamic seal.   (3 marks)
46 State one safety precaution related to each of the following:
    (a) Siting of an accumulator   (2 marks)
    (b) Accumulator charging equipment   (2 marks)
    (c) Removing an accumulator from a system.   (2 marks)
47 (a) Illustrate, using neat sketches, meter-in and meter-out flow control
      applications to control the outstroke.   (4 marks)
    (b) State which flow control method can be used with a load that tends
      to run away.   (2 marks)
48 Explain briefly how fluid leakage can be caused by:
    (a) A damaged piston rod   (2 marks)
    (b) Inadequately supported pipes   (2 marks)
    (c) Pressure shocks in the system.   (2 marks)
49 (a) State the primary function of an accumulator.   (1 marks)
    (b) Name the components A, B, C and D shown in Figure A2.18.   (4 marks)
Questions 50, 51 and 52 refer to Figure A2.19
50 Name and state the function of components A and B.   (6 marks)
51 (a) With Solenoid A energised and the main piston fully extended, state
      the pressure gauge readings at P1, P2 and P3.   (3 marks)
    (b) Name component C and state the main characteristics that would
      be checked when selecting a replacement for this component.   (3 marks)

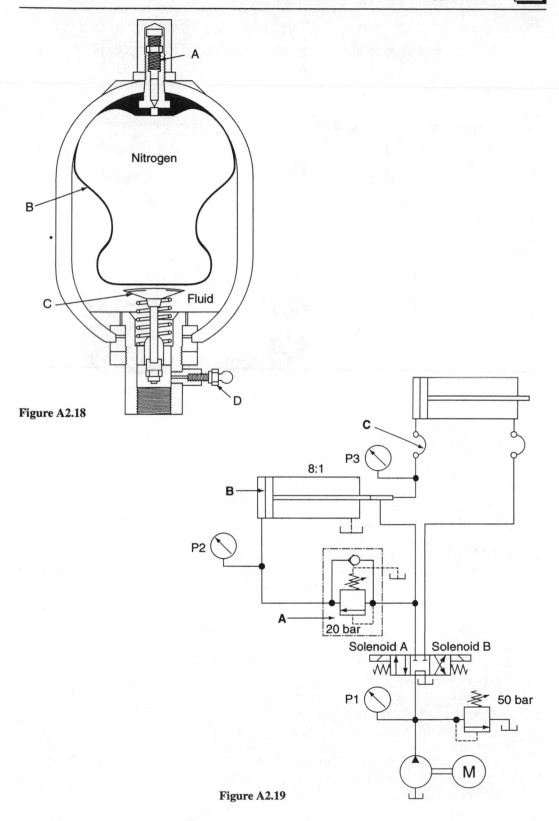

Figure A2.18

Figure A2.19

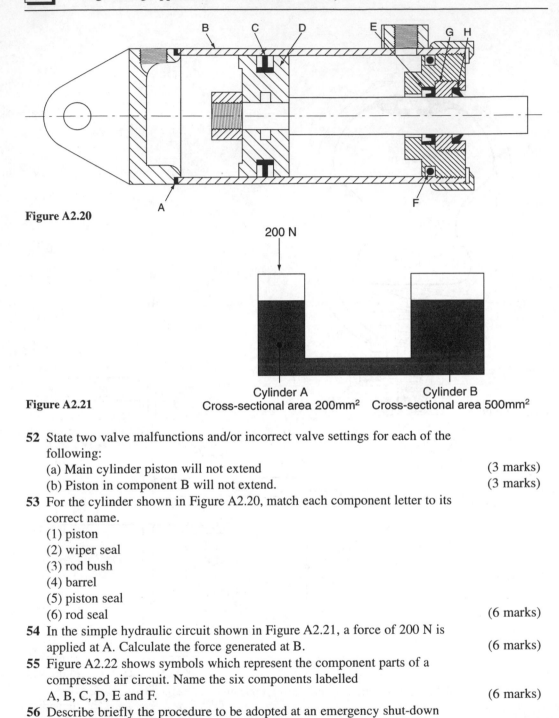

Figure A2.20

Figure A2.21

Cylinder A
Cross-sectional area 200mm²

Cylinder B
Cross-sectional area 500mm²

52 State two valve malfunctions and/or incorrect valve settings for each of the
following:
(a) Main cylinder piston will not extend (3 marks)
(b) Piston in component B will not extend. (3 marks)

53 For the cylinder shown in Figure A2.20, match each component letter to its
correct name.
(1) piston
(2) wiper seal
(3) rod bush
(4) barrel
(5) piston seal
(6) rod seal (6 marks)

54 In the simple hydraulic circuit shown in Figure A2.21, a force of 200 N is
applied at A. Calculate the force generated at B. (6 marks)

55 Figure A2.22 shows symbols which represent the component parts of a
compressed air circuit. Name the six components labelled
A, B, C, D, E and F. (6 marks)

56 Describe briefly the procedure to be adopted at an emergency shut-down
of a fluid power transmission system. (6 marks)

57 Make an outline diagram of an oil-mist lubrication system. (6 marks)

58 Describe the operation of an oil-mist lubrication system. (6 marks)

59 Figure A2.23 shows a hydraulic circuit. Describe, to BS 2917, the six
components labelled A, B, C, D, E and F. (6 marks)

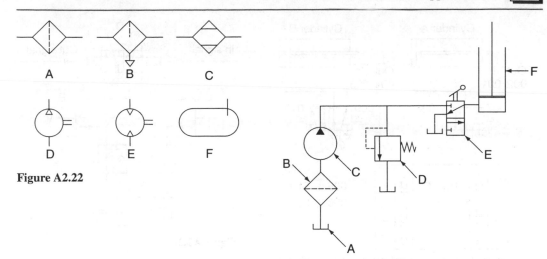

A　B　C

D　E　F

Figure A2.22

B　C　D　E　F

Figure A2.23

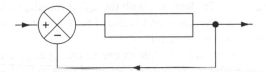

Figure A2.24

60 (a) State the purpose of control. (2 marks)
　　(b) Redraw Figure A2.24 and label clearly each of the following:
　　　　(i) desired input
　　　　(ii) comparator
　　　　(iii) process
　　　　(iv) feedback
　　　　(v) output. (4 marks)
61 State the purpose of each of the following:
　　(a) A control system actuator (3 marks)
　　(b) A process control 'signal processor'. (3 marks)
62 State the reason for preferring a hydro-pneumatic speed control system to
　　a purely pneumatic system. (6 marks)
63 A programmable logic controller (PLC) is used to sequence two cylinders.
　　Use Figure A2.25 and list:
　　(a) The input signals required to the PLC (3 marks)
　　(b) The output signals required from the PLC. (3 marks)
64 State two advantages of electrical control over a pneumatically signalled
　　system. (6 marks)
65 Electrical instrument transmission systems are not subject to transmission
　　distance and freezing temperatures as pneumatic systems. When using
　　pneumatic systems, state:
　　(a) the approximate limits on transmission distance (3 marks)
　　(b) the air supply precaution required at freezing temperatures. (3 marks)

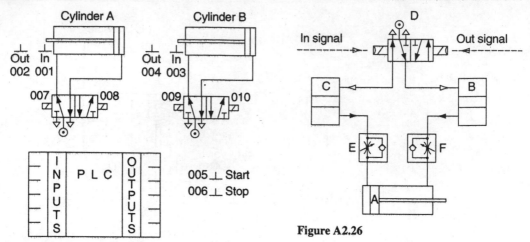

Operating devices are numbered 001 through to 010

Figure A2.25

Figure A2.26

66 With reference to Figure A2.26 describe, with the aid of the diagram, the
   action of the air/oil actuator system used to control the position of piston A.   (6 marks)
67 (a) List three types of proximity detector.   (3 marks)
   (b) Describe briefly the operating principle of one of the types given in (a).   (3 marks)
68 Describe two operating problems with direct solenoid-operated valves.   (6 marks)

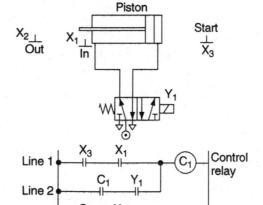

Figure A2.27

69 Figure A2.27 shows a simple ladder logic diagram for operating a piston.
   State the circuit function of each of the following:
   (a) X_1 line 1   (3 marks)
   (b) X_2 line 3.   (3 marks)

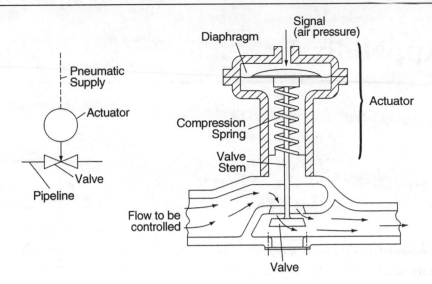

Figure A2.28

70 Refer to Figure A2.28.
   (a) Name the type of valve shown.   (2 marks)
   (b) State its flow characteristic.   (1 mark)
   (c) Describe the operation of the valve.   (3 marks)

Appendix 3

Institution of Plant Engineers

Relevant technical guides

- Compressed Air Systems
- Pipe Sizing – Hydraulic Transmission of Power With Uniform Flow
- Fault Finding Using Functional Charts
- Fault Diagnosis
- Programmable Logic Controllers.

The above guides can be obtained from:

The Secretary
The Institution of Plant Engineers
77 Great Peter Street
Westminster
London
SW1P 2EZ

Tel: 0171 233 2855
Fax: 0171 233 2604

Appendix 4

Standards and standardisation organisations

Relevant fluid power standards

ISO 5598	Hydraulic and pneumatic power vocabulary
ISO 1219 (BS 2917)	Fluid power systems and components graphical symbols
CETOP RP41	Hydraulic and pneumatic systems circuit diagrams
CETOP RP68P	Identification code for ports and operators of pneumatic control valves and other components
ISO 6431 and 6432	Mounting dimensions of single rod double-acting pneumatic cylinders 8 mm to 320 mm bore
ISO 5599	Pneumatic 5-port directional control valve. Mounting interface surface dimensions sub-base mountings
BS 4575 (Parts 2 and 3)	Code of practice for pneumatic equipment and systems
BS 1123	Safety valves, gauges and flexible plugs for compressed air or inert gas installations
BS 1780	Bourdon tube pressure and vacuum gauges
BS 4509	Methods of evaluating the performance of pneumatic transmitters with 3 to 15 psi (gauge) output
BS 4151	Recommended form of evaluating pneumatic valve positioners with 3 to 15 psi (gauge) input signal.

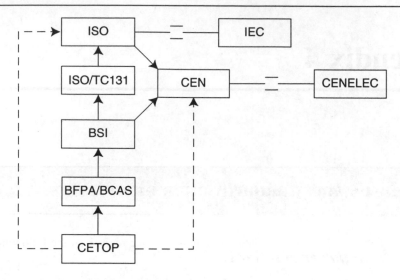

Figure A4.1 International organisation of fluid power standards

Some national and international standardisation organisations

ISO	International Standards Organisation
TC131	A fluid power committee within ISO
IEC	International Electro-Technical Commission
BSI	British Standards Institute
BFPA	British Fluid Power Association
BCAS	British Compressed Air Society
CEN	Comité Européen de Coordination des Norme (European Standards Committee)
CETOP	Comité Européen des Transmissions Oleohydrauliques et Pneumatiques (European Hydraulics & Pneumatics Standards Committee)
CENELEC	Comité Européen des Normalisation Electrotechnique (European Electrotechnical Standards Committee).

The national standards organisations in France, Germany and the UK are:

AFNOR	France
DIN	Germany
BSI	UK

Appendix 5

Bar litre – Calculations

The concept of bar litres may not be entirely familiar to some readers so the following illustrates some simple calculations:

Example 1

Calculate the bar-litres for the tank shown below which contains a fluid at a nominal pressure of 100 bar:

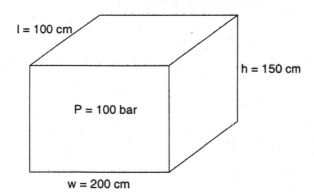

l = 100 cm
h = 150 cm
P = 100 bar
w = 200 cm

Steps for solution:

1) First find the volume of the tank in litres.
   Recall from page 6 that

   $1 \text{ m}^3 = 1000$ litres.

Now,

   $1 \text{ m} = 100 \text{ cm}.$

Therefore

   $1 \text{ m}^3 = 1\,000\,000 \text{ cm}^3 = 1000$ litres

and so

   $1 \text{ litre} = 1000 \text{ cm}^3.$

Hence, to find the volume in litres of the tank, take cm^3 and divide by 1000.

$$\text{Vol} = \frac{l \text{ (cm)} \times h \text{ (cm)} \times w \text{ (cm)}}{1000} \text{ in litres.}$$

Therefore,

$$\text{Vol} = \frac{100 \times 150 \times 200}{1000} = 3000 \text{ litres.}$$

2) Next multiply this result by the pressure in bar to obtain bar-litres.ie, bar-litres = bar × litres

$= 100 \times 3000$

$= 300\,000$ bar litres

Therefore, this vessel would be categorised as an '**Intermediate system**' under the pressure systems regulations since it exceeds 200 000 bar litres (minor system) but is below 1 000 000 bar litres (major system).

Example 2

The following storage cylinder contains a gas at 18 bar pressure. Determine the bar litre.

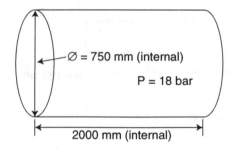

Internal diameter ∅ in metres

$$= \frac{750}{1000} = 0.75 \text{ m.}$$

Length in metres

$$= \frac{2000}{1000} = 2 \text{ m.}$$

Volume of cylinder

$$\pi R^2 l \text{ or } \frac{\pi D^2 l}{4.}$$

Therefore, volume

$$= \frac{\pi \times (0.75)^2 \times 2}{4} = 0.88 \text{ m}^3$$

Now recall that 1 m^3 = 1000 litres.

Therefore,

0.88 m^3 = 0.88 × 1000 = 880 litres.

So,

bar litres = 18 × 880 = 15 840 bar litres,

which is a **minor system**.

Index